T0214580

SpringerBriefs in Molecular Science

Biobased Polymers

Series editor

Patrick Navard, CNRS/Mines ParisTech, Sophia Antipolis, France

Published under the auspices of EPNOE*Springerbriefs in Biobased polymers covers all aspects of biobased polymer science, from the basis of this field starting from the living species in which they are synthetized (such as genetics, agronomy, plant biology) to the many applications they are used in (such as food, feed, engineering, construction, health, …) through to isolation and characterization, biosynthesis, biodegradation, chemical modifications, physical, chemical, mechanical and structural characterizations or biomimetic applications. All biobased polymers in all application sectors are welcome, either those produced in living species (like polysaccharides, proteins, lignin, …) or those that are rebuilt by chemists as in the case of many bioplastics.

Under the editorship of Patrick Navard and a panel of experts, the series will include contributions from many of the world's most authoritative biobased polymer scientists and professionals. Readers will gain an understanding of how given biobased polymers are made and what they can be used for. They will also be able to widen their knowledge and find new opportunities due to the multidisciplinary contributions.

This series is aimed at advanced undergraduates, academic and industrial researchers and professionals studying or using biobased polymers. Each brief will bear a general introduction enabling any reader to understand its topic.

*EPNOE The European Polysaccharide Network of Excellence (www.epnoe.eu) is a research and education network connecting academic, research institutions and companies focusing on polysaccharides and polysaccharide-related research and business.

More information about this series at http://www.springer.com/series/15056

Matej Bračič · Simona Strnad
Lidija Fras Zemljič

Bioactive Functionalisation of Silicones with Polysaccharides

Springer

Matej Bračič
Laboratory for Characterization
   and Processing of Polymers (LCPP),
   Faculty of Mechanical Engineering
University of Maribor
Maribor, Slovenia

Lidija Fras Zemljič
Laboratory for Characterization
   and Processing of Polymers (LCPP),
   Faculty of Mechanical Engineering
University of Maribor
Maribor, Slovenia

Simona Strnad
Laboratory for Characterization
   and Processing of Polymers (LCPP),
   Faculty of Mechanical Engineering
University of Maribor
Maribor, Slovenia

ISSN 2191-5407   ISSN 2191-5415   (electronic)
SpringerBriefs in Molecular Science
ISSN 2510-3407   ISSN 2510-3415   (electronic)
Biobased Polymers
ISBN 978-3-030-02274-7   ISBN 978-3-030-02275-4   (eBook)
https://doi.org/10.1007/978-3-030-02275-4

Library of Congress Control Number: 2018957640

This Springer imprint is published by the registered company Springer Nature Switzerland AG
The registered company address is: Gewerbestrasse 11, 6330 Cham, Switzerland

Contents

About the Authors

Dr. Matej Bračič is an expert in studying interactions of molecular and polymer dispersions in liquids with solid surfaces and biological evaluation of both surfaces. His research work is focused on the preparation and characterisation of nano and micro-dispersion, the preparation and characterisation of solid surfaces by electrospinning, spin-coating, dip-coating and dip-casting methods, interactions with biomolecules (proteins, enzymes, bacteria) and the transfer of knowledge into the field of sensors for the production of bioactive sensors. His bibliography covers 65 units, of which 21 are original scientific articles, 29 are contributions at national and international conferences and 14 are final reports on national and international projects. He has 91 pure citations in the last 10 years and an h index of 7. He received an award at the International Conference IN-TECH 2014: Award for Science and Technology Transfer (World Association of Innovative Technologies) on 12 September 2014 in Leiria, Portugal. He is currently working as a postdoctoral researcher at the Laboratory for Characterisation and Processing of Polymers, Institute of Engineering Materials and Design, University of Maribor.

Prof. Dr. Simona Strnad, Dr. Tech. Sci. is an expert in structure and properties of fibrous materials and advanced materials engineering. Her research work is focused on investigations of structure/properties relationships of polymer materials and surface functionalisation and characterisation. Her bibliography comprises 351 bibliographic items including 63 scientific papers, 7 Invited lectures, 6 chapters, 2 patents and more than 60 research and industrial project collaborations, among them FP6 NMP3-CT-2005-500375: EPNOE—Polysaccharides; Network of Excellence "Polysaccharides" (COBISS.SI-ID 13864470); FP7 NMP-2007-2.1.1-1—SURFUNCELL—Nanostructured polymer-matrix composites "Surface functionalisation of cellulose matrices using coatings of functionalised polysaccharides with embedded nano-particles"; EUREKA FeVal E!5851, EUREKA Vascucharge E! 2866, EUREKA Eurocharge E!2821; EUREKA Hightampons E!3602. Currently, she has a full professor position at the Institute of Engineering Materials and Design, University of Maribor.

Prof. Dr. Lidija Fras Zemljič (born 1973) received a Ph.D. in Chemistry at University Maribor in 2004 for her research on bio-functionalisation for achieving improved sorption properties. She further develops her skins and gain knowledge during her postdoctoral stay at Laboratory of "Fisica Aplicada", University of Granada, Spain and Laboratory for Forest Products Chemistry, Helsinki University of Technology, Finland during 2003–2004 years. She was/is a principal investigator and project manager in many national (ARSS agency) and industry-funded research projects (EURECA; COST) and collaborated in many FP6, FP7 and H2020 research and industrial project collaborations, among them FP6 NMP3-CT-2005-500375: EPNOE—Polysaccharides; Network of Excellence "Polysaccharides" (COBISS.SI-ID 13864470); FP7 NMP-2007-2.1.1-1—SURFUNCELL—Nanostructured polymer-matrix composites "Surface functionalisation of cellulose matrices using coatings of functionalised polysaccharides with embedded nano-particles", etc.

Her bibliography comprises 304 bibliographic items including 67 scientific papers, 5 Invited lectures, 7 chapters, 2 patents. Currently, she has a full professor position at the Institute of Engineering Materials and Design, University of Maribor.

Abbreviations

AAL	Allylalcohol
AAM	Allylamine
AFM	Atomic force microscopy
ALGA	Alginic acid
ASM	Active surface modification
BSA	Bovine serum albumin
CAC	Critical association constant
CAUTI	Catheter-associated urinary tract infection
CFU	Colony-forming units
Chi/CT	Chitosan
CLSM	Confocal laser scanning microscopy
CMC	Critical micelle concentration
CMChi/CMCT	Carboxymethyl chitosan
CVC	Cardiovascular catheter
FDA	Food and drug administration
FTIR	Fourier-transform infrared
GMK	Green monkey kidney
HA	Hyaluronic acid
HEMA	2-hydroxiethylmethacrylate
HSA	Human serum albumin
LCC	Local carriers or coatings
LDH	Lactate dehydrogenase activity
MIC	Minimum inhibitory concentration
MKM	Lysine-derived cationic surfactant
MTT	(3-(4,5-dimethylthiazol-2-yl)-2,5-diphenyltetrazolium bromide assay)
PBS	Phosphate buffer saline
PDMS	Polydimethyl siloxane
PEG	Polyethylene glycol
PEMA	Poly(ethylene-alt-maleic anhydride)

PET	Polyethylene terephthalate
PSM	Passive surface modification
PTFE	Polytetrafluoroethylene
PVC	Polyvinyl chloride
QCM-D	Quartz crystal microbalance with dissipation
RH	Relative air humidity
SEM	Scanning electron microscopy
TOS	Tosufloxacin
USA	United States of America
UTI	Urinary tract infection
UV	Ultraviolet
UVO	Ultraviolet ozone
XPS	X-ray photoelectron microscopy

Introduction

This book deals with the surface functionalisation of silicones, used as medical implants. Its main focus is on urethral catheters as medical implants. Surface functionalisation of urethral catheters has some specifics but also some general approaches; therefore, the described modifications can be transferred to other medical implants made of silicone as well.

The development of antimicrobial and antifouling surfaces for materials used as urethral catheters has been in the interest of the medical community for several decades as urinary tract infections (UTIs) have shown to be amongst the most prevalent ones having the highest occurrence of all infections in health institutions. Statistical data indicates that amongst newly submitted patients, 23% of all infections are UTI. 80% of which are catheter-associated UTI (CAUTI) [1]. Urethral catheters are being used by 5–10% of patients in long-term care, mostly incontinent males and females [2]. Incontinence is a major reason for catheterisation since it affects 50% of the elderly population and 25% of these receives catheter treatment during hospitalisation [1]. In the USA, 40% of all infections patients get during hospitalisation are UTI with a mortality rate of 3% [3]. It has been shown that the major reasons for the appearance of CAUTI are errors in catheter management, such as opening a closed drainage system or false handling during insertion and usage of the catheter [4]. The European program of CAUTI prevention, therefore, includes (i) education of the staff regarding the indication of dates of catheter insertion and removal, (ii) documentation of catheter insertion and use, (iii) the selection of appropriate catheter materials and (iv) its catheter maintenance [5].

Besides the health issues, curing CAUTI also has a considerable economic impact. The overall cost for medical treatment of UTI in USA is staggering, with around 450 million US$ spent annually to manage these infections. The cost for treatment of one episode of CAUTI is approximately 2900 US$ [6, 7]. The European Union is facing the same problems, from all the patients suffering from UTI, 63% are CAUTI, with a mortality rate of 1.8%. Most of which are elderly people. This problem will be even more pronounced in the future due to the fact that by 2050, it is projected that the global population aged 65 years and above will triple.

The presence of an indwelling urethral catheter bypasses the normal defence mechanisms of the host and allows microorganisms' continuous access to the urinary bladder. The microorganisms can ascend on the inner or outer surface of the catheter and persist once they reach the bladder [2, 3]. The presence of pathogen bacteria in the urine is referred to as bacteriuria and can lead to life-threatening illnesses and even death [4]. The number of viable bacteria in the urine during bacteriuria is over 105 CFU/mL [3]. The majority are faecal contaminants or skin residents from the patient's own microflora that colonise the periurethral area [6]. In 3–7% of the patients, new UTI causing microorganisms to appear in the urine daily. With such a trend, the prevalence reaches 100% in 30 days of catheter use. The prevalence is a statistical quantity used in the field of epidemiology describing the disease frequency of a statistical population in a certain time period. The use of an indwelling urethral catheter increases the prevalence for UTI for thirty times when compared to UTI prevalence for patients without urethral catheter [2]. The normal defence mechanism of the human host when bacteria bind to the bladder mucosa is inflammation that results in an influx of neutrophils and sloughing of epithelial cells. Both processes contribute to clearance of the bacteria from the mucosal surface. The urethral catheter does not possess such inherent defence mechanisms and bacteria can easily colonise its surface leading to the formation of a biofilm which causes additional problems in the process of preventing and curing of CAUTI [8].

The first step in biofilm formation on a urethral catheter is deposition of a conditioning film of host urinary components, including proteins, electrolytes, and other organic molecules [8]. Free-flowing bacteria attach to the surface through hydrophobic and electrostatic interactions and through the use of flagella where they form a self-organised cooperative community with high resistance to antimicrobial treatment [8, 9]. Intracellular communication by quorum sensing regulates the formation of such three-dimensional communities forming a biofilm which cannot be removed by simple shear forces and offers good growing conditions for other free-flowing bacteria. This leads to clogging of the catheter lumen and promotes bacterial colonisation. The resistance of bacteria inhabiting the biofilm can be up to thousand times higher when compared to free-flowing bacteria in urine [6, 9, 10]. In addition to biofilm clogging, the bacteria *Proteus mirabilis* causes clogging and encrustation of the catheter by precipitation and adsorption of calcium and magnesium phosphate present in urine. Long-term use of indwelling catheters results in salt encrustation in 50% of the cases [11, 12]. In some cases, an excess of proteins can also be present in the urine which additionally contributes to fouling of the urethral catheter. The presence of excess protein in urine is referred to as proteinuria. So far, there is no evidence that would directly link UTI as the cause for proteinuria [13], but many patients who need catheterisation develop proteinuria due to the nature of their medical state, such as people who suffer from spinal cord injuries [14]. Because of the latter and the role of proteins in the biofilm formation, their presence must be taken into consideration when developing antimicrobial and antifouling urethral catheters.

Overcoming these problems is the driving force for the development of surface-modified bioactive urethral catheters. New trends, including coatings specifically targeting bacterial deposition, natural polymer coatings, enzyme active coatings or biomolecular coatings, are slowly replacing conventional antibiotic treatments and metal ion coatings [15].

References

1. J. Elvy, A. Colville, Catheter associated urinary tract infection: what is it, what causes it and how can we prevent it? J. Infect. Prev. **10**, 36–41 (2009). https://doi.org/10.1177/1757177408094852
2. L.E. Nicolle, The chronic indwelling catheter and urinary infection in long-term-care facility residents. Infect. Control Hosp. Epidemiol. **22**, 316–321 (2001). https://doi.org/10.1086/501908
3. K. Schumm, T.B.L. Lam, Types of urethral catheters for management of short-term voiding problems in hospitalised adults. Cochrane Database Syst. Rev., 110–121 (2008). https://doi.org/10.1002/14651858.cd004013.pub3
4. P.A. Pinzon-Arango, Y. Liu, T.A. Camesano, Cranberry juice components act against development of urinary tract infections. Bioeng. Proc. Northeast Conf., (2009). https://doi.org/10.1109/nebc.2009.4967636
5. B.W. Trautner, Management of Catheter-Associated Urinary Tract Infections (CAUTIs). Virology **23**, 76–82 (2010). https://doi.org/10.1097/qco.0b013e328334dda8.management
6. S.M. Jacobsen, D.J. Stickler, H.L.T. Mobley, M.E. Shirtliff, Complicated catheter-associated urinary tract infections due to escherichia coli and proteus mirabilis. Clin. Microbiol. Rev. **21**, 26–59 (2008). https://doi.org/10.1128/cmr.00019-07
7. U.-S. Ha, Y.-H. Cho, Catheter-associated urinary tract infections: new aspects of novel urinary catheters. Int. J. Antimicrob. Agents **28**, 485–490 (2006). https://doi.org/10.1016/j.ijantimicag.2006.08.020
8. B. Trautner, R. Darouiche, Role of biofilm in catheter-associated urinary tract infection. Am. J. Infect. Control **32**, 177–183 (2004). https://doi.org/10.1016/j.ajic.2003.08.005
9. L. Muzzi-Bjornson, L. Macera, Preventing infection in elders with long-term indwelling urinary catheters, J. Am. Acad. Nurse Pract. **23**, 127–134 (2011). https://doi.org/10.1111/j.1745-7599.2010.00588.x
10. D.R. Monteiro, L.F. Gorup, A.S. Takamiya, A.C. Ruvollo-Filho, E.R. de Camargo, D.B. Barbosa, The growing importance of materials that prevent microbial adhesion: antimicrobial effect of medical devices containing silver. Int. J. Antimicrob. Agents **34**, 103–110 (2009). https://doi.org/10.1016/j.ijantimicag.2009.01.017
11. E.L. Lawrence, I.G. Turner, Materials for urinary catheters: a review of their history and development in the UK. Med. Eng. Phys. **27**, 443–453 (2005). https://doi.org/10.1016/j.medengphy.2004.12.013
12. D.J. Stickler, Bacterial biofilms in patients with indwelling urinary catheters. Nat. Clin. Pract. Urol. **5**, 598–608 (2008). https://doi.org/10.1038/ncpuro1231

13. J.L. Carter, C.R.V. Tomson, P.E. Stevens, E.J. Lamb, Does urinary tract infection cause proteinuria or microalbuminuria? A systematic review. Nephrol. Dial. Transplant. **21**, 3031–3037 (2006). https://doi.org/10.1093/ndt/gfl373

14. S. Vaidyanathan, K.A. Abraham, G. Singh, B. Soni, P. Hughes, Screening for proteinuria in "at-risk" patients with spinal cord injuries: lessons learnt from failure. Patient Saf. Surg. **8**, 1–8 (2014). https://doi.org/10.1186/1754-9493-8-25

15. A. Vaterrodt, B. Thallinger, K. Daumann, D. Koch, G.M. Guebitz, M. Ulbricht, Antifouling and antibacterial multifunctional polyzwitterion/enzyme coating on silicone catheter material prepared by electrostatic layer-by-layer assembly. Langmuir **32**, 1347–1359 (2016). https://doi.org/10.1021/acs.langmuir.5b04303

Chapter 1
Silicone in Medical Applications

1.1 Structure and Properties of Silicone

Silicones are oligomers or polymers of organic siloxanes, which are products of organochlorosilanes [1]. The basis for all silicones is quartz sand or silica from which silicon is obtained in the first phase (Eq. 1.1). In the coming phases, silicone is obtained from siloxanes or silanes. The reaction of silicone production from silicon dioxide is conducted at 1700 °C [2, 3].

$$SiO_2 + C \rightarrow Si + CO_2 \tag{1.1}$$

Silanes are silicon molecules to which hydrogen ions are bound. The simplest representative of silanes is mono silane, which has four hydrogen ions bound to one silicon atom. Bonds between silcon and carbon are formed in the presence of methylene chloride and a catalyst, resulting in the formation of organosilanes. Organosilanes are silanes which have at least one of the hydrogen atoms substituted with an organic substituent. The synthesis of organosilanes usually is not conducted directly from silanes, but from silane halides in which halogen atoms are bound to the silicon atom instead of hydrogen ones. If both organic substituents as well as halides are bound to the silicon atom one speaks of organosilane halides, which are the basis for silicone polymer synthesis. The main representative of organosilane halides are alkylsilane halides. An example of dimethyldichloro silane synthesis is shown in Eq. (1.2) [2, 3].

$$2CH_3-Cl + Si \rightarrow (CH_3)_2-Si-Cl_2 \tag{1.2}$$

Dialkyldichloro silanes have the ability to hydrolyse and form dialkyl silanols (Eq. 1.3), which are able to polycondensate and form silicone polymers linked together with the Si–O–Si siloxane bond [2, 3].

© The Author(s), under exclusive licence to Springer Nature Switzerland AG 2018
M. Bračič et al., *Bioactive Functionalisation of Silicones with Polysaccharides*,
Biobased Polymers, https://doi.org/10.1007/978-3-030-02275-4_1

$$(CH_3)_2-Si-Cl_2 + 2H_2O \rightarrow 2HCl + (CH_3)_2-Si-(OH)_2 \qquad (1.3)$$

Basically, all silicone elastomers are based on the $-Si-O-$ bond. An example of the siloxane chain formation from trimethylsilanol is shown in Eq. (1.4) [2–4].

$$2(CH_3)_3-Si-OH \rightarrow (CH_3)_3-Si-O-Si(CH_3)_3 + H_2O \qquad (1.4)$$

The most common and most basic representative of siloxane polymers is poly-dimethyl siloxane (PDMS). Crosslinking of PDMS is shown in Fig. 1.1.

Silicone elastomers used for the manufacturing of urethral catheters are usually produced by (i) crosslinking of polysiloxane macromolecules by radical reactions (if vinyl functions are present in the polymer chain), (ii) by condensation, (iii) by reactions of addition, and (iv) by incorporation of fillers, which act as reinforcement agents [3, 4].

The silicone elastomer exhibits excellent biocompatible properties, low toxicity and low tissue inflammation degree [5]. Some physicochemical properties of the silicone elastomer are shown and compared with natural latex in Table 1.1. All types of silicones are chemically and physiologically inert, resistant to extreme temperature conditions, and have excellent electrical insulation and antiadhesive properties. The low adhesion degree is a result of silicones' low surface free energy of 24 mJm^{-2} and the low cohesive strength of the most outer surface layer [6]. Nevertheless, non-specific protein adsorption is often observed when silicone-based implants are used in contact with human tissue due to the hydrophobic nature of PDMS, enabling hydrophobic interactions with proteins from blood and other body fluids [7–9]. Additional introduction of antifouling properties is therefore significant when considering the use of silicone elastomers as medical implants e.g. urethral catheters.

As can be seen in Table 1.1, silicones have slightly lower tensile strength values than natural latex, but on the other hand exhibit superior resistance to UV radiation, chemical action, and do not cause allergic reactions when in contact with human tissue. The latter is extremely important when choosing the appropriate material for short and long-term catheter manufacturing [10]. The higher rigidness of silicone

Fig. 1.1 **a** Chemical structure of PDMS. **b** An example of crosslinking of PDMS by polycondensation

Table 1.1 Mechanical and chemical properties comparison of natural latex and the silicone elastomer [5]

Properties	Silicone	Latex
Tensile strength (MPa)	2.4–7	7–30
Elongation (%)	350–600	100–700
UV-resistance	Excellent	Poor
Chemical resistance	Excellent	Poor
Allergic reaction	No	Yes

allows one to produce catheters with very thin walls and consequently wider lumens. This increases the urine flow and it has been shown that catheter blockage appears later in silicone catheters than in the case of catheters made of other materials [5].

Once the polysiloxane molecules polycondensate and form crosslinked networks they can be, in dependence of the used crosslinker, oriented linearly, cyclic, brushed, star-shaped, or randomly [11]. Regardless of the shape, they all show low intermolecular forces between the methyl groups, high chain flexibility due to high bonding energy between silicon and oxygen, and a partially ionic nature of the siloxane bond (Table 1.2) [12].

The cross-linked siloxane structures in the silicone elastomer can be crystalline or amorphous. The crystallinity degree depends on the temperature during the heating phase in the curing process and during the cooling phase after the curing process has completed. The polysiloxane macromolecules have enough time for spatial rearrangement when the elastomer is cooled down slowly after the curing process, which results in high crystallinity. On the other hand, fast cooling leads to elastomers that are more amorphous. By optimising the heating and cooling parameters, elastomers with a crystallinity degree of up to 85% can be prepared [11]. The velocity of the cooling influences the crystal formation also in other ways. Fast cooling maintains only the crystals formed during heat treatment while slow cooling slow cooling enables macromolecules to rearrange and form larger crystalline regions with less defects. In Table 1.3 some data on silicone elastomer crystal cell dimensions are shown. The average molecular weight of PDMS macromolecules also has a significant impact on the formation of crystals, besides the temperature during curing and cooling process. Shorter molecules crystallise at lower temperatures and form more perfect crystals due to their higher mobility in comparison to longer molecules. One should also not neglect the influence of the molecular orientation, as described in the beginning of this section, regarding crystal formation. As for most polymers, Linear macromolecules form elastomers with a much higher crystallinity degree when compared to elastomers formed from randomly branched macromolecules [11].

Table 1.2 Chemical properties of PDMS [5]

Properties	Value
Chain rotation energy (kJ/mol)	≈0
Siloxane bond energy (kJ/mol)	445

Table 1.3 Crystal cell dimensions of the silicone elastomer from PDMS [13]

Number of monomers per unit cell	Cell dimension (Å)			Cell angles			Density (gcm^{-3})	
	a	b	c	α	β	γ	Crystalline	Amorphous
6	13.0	8.3	7.8	90°	60°	90°	1.07	0.98

The monoclinic unit cell of the silicone elastomer from PDMS as determined by Damaschun and co-authors is shown in Fig. 1.2.

A very special feature of silicone elastomers are their stable mechanical properties which are retained even after forty years of aging. A Rapra technology LT, report from the year 2000, shows the mechanical properties of rubber materials after forty years aging. The resulted properties of aged natural latex and silicone are presented in Table 1.4 [15].

As can be seen from Table 1.4, the mechanical properties of latex are superior to the silicone ones but are reduced for 50% after 40 years of aging, while the ones of silicone elastomer remain constant throughout the whole-time span. These results give good insides on the stability of the silicone elastomer, which is important when considering its long-term use.

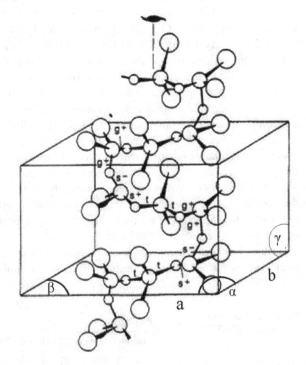

Fig. 1.2 Monoclinic unit cell of crystalline PDMS according to Damaschun. Large circles represent methyl groups, middle ones silicon atoms, and small ones represent oxygen atoms [14]. The figure was reprinted with permission from Schilling et al. [14]. Copyright Nov 1, 1991, American Chemical Society

Table 1.4 Mechanical properties of new and 40 years aged natural latex rubber and silicone elastomer [15]

Properties	New latex	Aged latex[a]	New silicone	Aged silicone[a]
Tensile strength (MPa)	25.3	16.5	7.07	8.23
Elongation at break (%)	428	223	113	121
Modulus at 100% elongation	4.42	6.38	7.13	6.54

[a]The forty years aging process was performed in real time, at a temperature interval of $T = [18, 25]$ °C, and a relative air humidity of $RH = 50 \pm 10\%$

1.2   Applications of Silicones in Medicine

The presence of organic groups attached to the inorganic backbone of silicones gives them unique properties and makes them useful in a wide variety of technological fields. Silicone is used in the electrical field as an insulating material, potting compound and in other applications specific for semiconductor manufacturing [12]. It is used in aerospace industry due to its excellent performance in low and high temperature conditions. Its durability makes it a perfect material for sealants, adhesives, and water proof coatings in the construction industry [16]. But due to its exceptional biocompatibility silicone is labelled as one of the most promising materials for the manufacturing of medical implants [3, 12, 17].

It was in the mid-1940s when scientists at the University of Toronto found out that rinsing methylchlorosilane coated needles, syringes, and vials with distilled water results in formation of a silicone coating which can be used for blood coagulation prevention [3]. Up to this day, most needles and syringes are still coated with silicone, but the material has over the course of the years paved its way into a diverse spectrum of medical applications [3, 16]. Its biocompatibility in combination with chemical stability and its elastic nature makes it an ideal material for long-term implantation [3]. Implants vary from bile duct repair, which was first conducted using silicone in 1946 to various shunts and valves from which the most acclaimed one was the ventriculo-atrial valve known as the Spitz-Holter valve, which was designed in 1956 by a father whose son was dying from hydrocephalus. The valve is still used today in an almost unchanged form [3]. After that, the possibilities for silicone use in medical applications seemed unlimited and it has ever since been used in orthopedics, various catheters, drains and shunts, contact lenses, artificial organs, as components in kidney dialysis, heart-bypass machines, blood-oxygenators, and many others [3, 18]. They have almost completely replaced other materials, such as polyvinyl chloride (PVC) or latex for manufacturing of urethral catheters. Silicone catheter raw materials are more expensive than latex and PVC, but when examining healthcare choices, other factors should be taken into account. Curtis and Klykken [19] reported that catheters produced from silicone mostly improve patient comfort and reduce total patient cost by reducing the occurrence of allergic responses, incidence of phlebitis, frequency of sepsis, number of catheter insertions, likelihood of mineral encrustations, potential for bacterial migration, occurrence of premature balloon deflation, and potential for nosocomial infections [19].

Silicones, are one of the most thoroughly tested and widely-used groups of bio-materials for catheter manufacturing, due to their intrinsic biocompatibility and bio-durability. However, as pointed out the catheter associated urinary tract infections still represent a serious problem that needs to be solved. A basic strategy to deal with the above described problems of CAUTI and biofilm formation is aimed towards the modification of silicone surface properties i.e. introduction of a coating with a desired agent, and by manipulation of the surface roughness or morphology which can prevent the attachment of bacteria to the silicone catheter [20].

1.3   Urethral Catheters

Urethral catheters are used for urine secretion in patients with urine blockage and patients suffering from incontinence, as well as during surgery when normal urine secretion is disabled [21]. They are traditionally made of man-made polymers such as PVC, polytetrafluoroethylene (PTFE), latex, and silicone [5, 10, 22]. PVC catheters are generally used for short-term catheterization and are considered as disposables, because their biocompatible properties are lower when compared to latex, silicone or PTFE. Long-term use of PVC indwelling catheters causes urethra infections due to the release of phthalates and other plasticizers [19]. PTFE catheters have better biocompatible properties than latex ones and also have a lower friction coefficient, which reduces irritation during catheter insertion. On the other hand, the toxicity of PTFE catheters is higher. The antifouling properties of PTFE catheters are compa-rable with latex catheters. Silicone catheters show the highest biocompatible prop-erties and the lowest toxicity of all the above-mentioned materials and are therefore the most promising ones when developing long-term indwelling urethral catheters. Their physicochemical properties allow one to produce catheters with a very thin wall, which increases the catheter lumen and reduces the possibility for protein and salt clogging. Silicone does not cause allergic reactions as latex can, but its drawback when compared to latex is its rigidity, which reduces the comfort during use [5, 23, 24]. The most common catheter used worldwide is the Foley catheter which was first made of latex and is now gradually replaced with silicone, mostly due to latex allergies that patients develop during use. Its scheme is shown in Fig. 1.3 [5, 10, 22].

The tip of the catheter, the drainage eye, and the deflated balloon are inserted in the urethra. Once inserted, the balloon is inflated through the inflation valve, which stabilizes the catheter and keeps it in place. The urine enters the lumen of the catheter through the drainage eye, runs down the catheter tube towards the drainage valve to which a collection bag is attached to collect the urine [5]. The diameter of the catheter's external lumen can vary from 1 to 11.3 mm depending on its application. The French scale is most commonly used to express the size of the catheters external

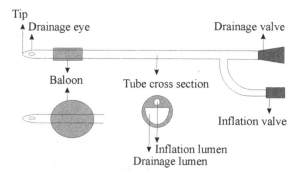

Fig. 1.3 Scheme of a conventional Foley catheter [25]

lumen in the unit French, which is equivalent to one third of a millimeter (Fig. 1.4) [26]. The catheters diameter plays an important role in preventing blockage due to biofilm formation on its surface. One should choose the appropriate diameter depending on the time period of the catheter usage. A large number of catheter types beside the Foley catheter are used in medical facilities and are generally, by the duration of their usage, divided in three major groups:

- disposable indwelling urethral catheters,
- short-term indwelling urethral catheters, and
- long-term indwelling urethral catheters.

Short-term catheters are used up to 30 days, while long-term catheters are used for time periods of more than 30 days. It is highly unlikely to avoid catheter blockage by choosing catheters of a larger diameter in the latter case so it is therefore a necessity to either shorten the time of usage or improve the antifouling properties of the catheter [5, 21]. Most short-term as well as long-term catheters are nowadays made of silicone or latex, because they show superior biocompatibility and durability to other man-made materials mentioned above. The material requirements for disposable catheters on the other hand, which are often used for self-catheterisation, are less strict in

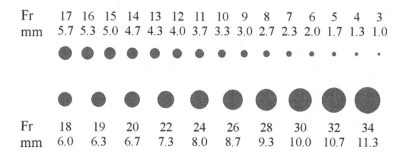

Fig. 1.4 Different catheter external lumen sizes expressed in French units and millimeters on the French scale [25]

comparison to short-term and long-term indwelling catheters and are therefore mostly made of PVC, which is a relatively inexpensive raw material when compared to silicone, PTFE or latex [5].

References

1. J. Navodnik, M. Kopčič, *Plastik-orodjar: priročnik* (Navodnik, Velenje, 1990)
2. A.J.O. Lenick, Basic silicone chemistry—a review. J. Surfactants Deterg. **3**, 387–393 (2000)
3. B.D. Ratner, *Biomaterials Science: An Introduction to Materials in Medicine* (Academic Press, San Diego, 1996)
4. E.G. Rochow, *An Introduction to the Chemistry of Silicones* (Wiley, New York, 1946)
5. E.L. Lawrence, I.G. Turner, Materials for urinary catheters: a review of their history and development in the UK. Med. Eng. Phys. **27**, 443–453 (2005). https://doi.org/10.1016/j.medengphy.2004.12.013
6. B. Crowther, *Handbook of Rubber Bonding* (2001)
7. J.G. Alauzun, S. Young, R. D'Souza, L. Liu, M.A. Brook, H.D. Sheardown, Biocompatible, hyaluronic acid modified silicone elastomers. Biomaterials **31**, 3471–3478 (2010). https://doi.org/10.1016/j.biomaterials.2010.01.069
8. L. Yang, L. Li, Q. Tu, L. Ren, Y. Zhang, X. Wang, Z. Zhang, W. Liu, L. Xin, J. Wang, Photocatalyzed surface modification of poly(dimethylsiloxane) with polysaccharides and assay of their protein adsorption and cytocompatibility. Anal. Chem. **82**, 6430–6439 (2010). https://doi.org/10.1021/ac100544x
9. A.J. Keefe, N.D. Brault, S. Jiang, *Using a Superhydrophilic Zwitterionic Polymer* (2012)
10. K. Schumm, T.B.L. Lam, Types of urethral catheters for management of short-term voiding problems in hospitalised adults. Cochrane Database Syst. Rev. 110–121 (2008). https://doi.org/10.1002/14651858.cd004013.pub3
11. S.J. Clarson, *Advanced Materials Containing the Siloxane* (2010), pp. 3–10
12. M.J. Owen, Properties and applications of silicones applications of PDMS. Silicones Silicone-Modified Mater. 13–18 (2010)
13. C.M. Kuo, Poly(dimethylsiloxane). Polym. Data Handb. 411–435 (1999). https://doi.org/10.1021/ja907879q
14. F.C. Schilling, M.A. Gomez, A.E. Tonelli, Solid-state NMR observations of the crystalline conformation of poly(dimethylsiloxane). Macromolecules **24**, 6552–6553 (1991)
15. R.P. Brown, T. Butler, *Natural Ageing of Rubber* (Birmingham, 2000)
16. M.J. Owen, P.R. Dvornic (eds.), *Silicone Surface Science* (Springer, 2012)
17. G. Bellussi, M. Bohnet, J. Bus, K. Drauz, H. Greim, K.-P. Jäckel, U. Karst, A. Kleemann, G. Kreysa, T. Laird, W. Meier, E. Ottow, M. Röper, K. Scholtz, J. Sundmacher, R. Ulber, U. Wietelmann, *Ullmann's Encyclopedia of Industrial Chemistry*, 7th edn. (Wiley-VCH, 2011)
18. A. Colas, J. Curtis, *Silicone Biomaterials : History and Chemistry Medical Applications of Silicones Dow Corning Corporation Biomaterials Science*, 2nd edn. About the Authors, Burns (2004), p. 20. https://doi.org/10.1016/b978-0-08-087780-8.00025-5
19. J. Curtis, P. Klykken, *A Comparative Assessment of Three Common Catheter Materials* (Dowcorningcom, 2008), pp. 1–8
20. B. Trautner, R. Darouiche, Role of biofilm in catheter-associated urinary tract infection. Am. J. Infect. Control **32**, 177–183 (2004). https://doi.org/10.1016/j.ajic.2003.08.005.Role
21. L.E. Nicolle, The chronic indwelling catheter and urinary infection in long-term-care facility residents. Infect. Control Hosp. Epidemiol. **22**, 316–321 (2001). https://doi.org/10.1086/501908
22. L. Muzzi-Bjornson, L. Macera, Preventing infection in elders with long-term indwelling urinary catheters. J. Am. Acad. Nurse Pract. **23**, 127–134 (2011). https://doi.org/10.1111/j.1745-7599.2010.00588.x

23. D.J. Chauvel-Lebret, P. Pellen-Mussi, P. Auroy, M. Bonnaure-Mallet, Evaluation of the in vitro biocompatibility of various elastomers. Biomaterials **20**, 291–299 (1999). https://doi.org/10.1 016/S0142-9612(98)00181-1
24. J. Park, R.S. Lakes (eds.), *Biomaterials: An Introduction* (Springer, 2000)
25. M. Bračič, *Surface Modification of Silicone with Polysaccharides for the Development of Antimicrobial Urethral Catheters* (Maribor, 2016)
26. K.V. Iserson, J.-F.-B. Charrière: the man behind the "French" gauge. J. Emerg. Med. **5**, 545–548 (1987)

Chapter 2
Catheter Associated Urethral Tract Infections

Urethral catheters should satisfy variety of requirements connected to the specific application. Besides antifouling and antimicrobial properties, they should meet optimal mechanical properties, which are mostly related to the comfort during use. The comfort is extremely important since it is a direct indication of possible damage to the urethral mucosa. A catheter should have a smooth surface bearing a low friction coefficient reducing mechanical irritation [1]. The relation between the flexibility and the strength of the material is also important. The more flexible the catheter is the less possible are damages to the urethral mucosa. On the other hand, lacking strength can cause destruction of the catheter during removal leading to mucosa damage as well [1, 2]. However, the importance of the relation is pronounced the most during balloon inflation. The balloon must be easily inflatable but should not burst during the process. Bursting balloons cause extreme bladder damage [3]. When speaking of basic material properties of a urethral catheter, the most important one is the biocompatibility of the chosen material. A biocompatible material should not harm the human body and should not cause an immune response. The degree of biocompatibility is determined by a set of standards from the ISO 10993 series, which regulate the methods of biocompatibility evaluation, toxicity tests of the material in contact with a living organism, and testing of biocompatibility in dependence of the material type and its degradation products [4].

2.1 The Role of Biofilm Formation

The microbial biofilm-associated infections caused by medical devices' implantation are still a major healthcare related complication (affect over 600 million patients worldwide annually, with approx. 4.1 million patients in Europe) that may result in serious discomfort and illness and, in many cases, even in the death of the patient [5]. Amongst them, UTI with a 23% share are the most frequent, followed by the orthopaedic endo-prosthesis infections. Amongst all UTI infections, 80% are CAUTI,

M. Bračič et al., *Bioactive Functionalisation of Silicones with Polysaccharides*,
Biobased Polymers, https://doi.org/10.1007/978-3-030-02275-4_2

which are affecting 10–25% of patients in long-term hospital care with an overall cost of around 395 million EUR per year and 2.550 EUR per treatment.

Besides catheter blockage the first problems usually occur immediately after its insertion. The catheter bypasses natural defence mechanisms and allows microorganisms to enter the urethra and the bladder, which results in UTI [6]. The most common microorganisms inhabiting the urine during a UTI are *Escherichia coli, Candida* spp., *Pseudomonas aeruginosa, Enterococcus* spp., *Klebsiella* spp., *Proteus* spp., *Enterobacter* spp., *Streptococcus* spp., *Citrobacter* spp., and *Staphylococcus aureus* [6–9]. The urinary catheter increases the chance of a UTI by mechanical irritation, stress of bladder due to the inflatable balloon, retention of urine in the bladder, and the physical damage of the bladder mucosa due to the presence of a catheter [10]. The entrance of bacteria into the urethra and their path to the bladder as well as the biofilm formation on the catheter's surface are schematically represented in Fig. 2.1.

In the right image of Fig. 2.1 one can see the position of a urethral catheter fixed in the bladder by the inflated balloon. The catheter enables microorganisms to enter the urethra and travel upwards to the bladder. These free-flowing pathogen organisms cause CAUTI and also play a major role in biofilm formation (Fig. 2.1, middle) [12]. The biofilm formation is extremely dangerous for the patient since it not only causes catheter blockage but also protects the pathogens inhabiting it from the host defence

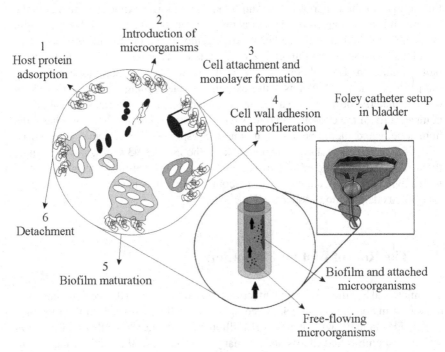

Fig. 2.1 A schematic representation of a Foley catheter setup in the human bladder (right), the presence of free-flowing and attached bacteria, and a step by step formation of a biofilm on the catheters surface [11]

mechanisms as well as antibiotic treatments, making it enormously difficult to fight CAUTI. A detailed step-by-step scheme of the biofilm formation is shown in the left image in Fig. 2.1. The first and very important step in biofilm formation is the adsorption of host proteins, electrolytes, and other organic molecules to the surface of the catheter once it has been inserted into the urethra [12]. An improved antifouling surface of the catheter could significantly reduce biofilm formation in this step already. What follows is the introduction of the host uropathogen microorganisms, which are mostly faecal contaminants or skin residents that colonise the periurethral area and are listed in the paragraph above [7, 10, 12]. Approximately two thirds of the pathogens enter the urethra on the external surface of the catheter and one third on the internal one [12]. The catheter bypasses natural host defences and allows microorganisms not only to enter the urethra but to grow fast and migrate to the bladder. Once the proteins cover the catheter's surface, host cells bind to them and form monolayers in the third step. The host cell receptors are recognised by the bacterial adhesins, which initiate the attachment of the pathogens. The adhesins have the ability to overcome electrostatic repulsion between the bacterial cell and the surface by recognising specific host cell surface and extracellular matrix components such as proteins, glycoproteins, glycolipids, and carbohydrates [12]. The adhesins can be found in the bacterial fimbria which are similar to flagella only much thinner and shorter. A vast variety of adhesins are produced by bacteria and each of them differs to the other. Their structure is not yet well defined but consists largely of proteins and carbohydrates, which once again emphasises the significance of the antifouling properties of the catheters surface. Once the pathogens are firmly attached to the catheter surface they begin to change phenotypically. They produce exopolysaccharides that cover and protect them allowing them to replicate and form microcolonies, which leads to biofilm maturation in the fourth step [12, 13]. The formation of three dimensional structures during maturation and the later detachment are regulated by quorum sensing [7, 12, 14]. In the fifth step, the detachment phase, shedding of the daughter cells from actively growing cells and the shearing of the biofilm aggregates seed other sections of the catheter surface [12]. If not properly treated, the whole process ultimately leads to severe health damage due to CAUTI and catheter blockage by the formation of biofilm [12]. Once a biofilm has developed on the inner or outer surface of a urinary catheter, the most useful way to eliminate the risk of CAUTI is to administrate antibiotics and remove the catheter [12]. Although, Gentamicin, minocycline, rifampicin, etc. [15, 16] were shown to significantly reduce the rate of gram-negative and gram positive bacteria, there are several problems related to this strategy. First, the pressing clinical question is the duration of the antibiotics treatment necessary to treat CAUTI. Second, due to the antibiotic use, bacterial resistance is developed and thus healing with antibiotics is limited. Various other prevention strategies generally fall under the headings of different types of catheters, different catheter materials, or alternatives to indwelling urinary catheters. One of the most attractive alternatives to indwelling urinary catheters is to provide antimicrobial, hydrophilic, and biofilm inhibiting surface properties [17]. Even more, despite certain promising results, the success rates of the most coatings are still highly variable; some of them may be toxic for the patient and environmentally problematic. Novel antimicrobial peptides are capable of avoiding this complication; however, some may still be potentially toxic for the host [12]. Therefore, the scientific and industrial spheres are exploring new possibilities for the improvement of antimicrobial and anti-biofilm-forming properties of catheters surface by using natural and biocompatible agents such as polysaccharides, natural surfactant and natural extracts [18, 19].

References

1. E.L. Lawrence, I.G. Turner, Materials for urinary catheters: a review of their history and development in the UK. Med. Eng. Phys. **27**, 443–453 (2005). https://doi.org/10.1016/j.mede ngphy.2004.12.013
2. M. Jenkins, A. Stamboulis (eds.), *Durability and Reliability of Medical Polymers* (Woodhead Publishing Limited, 2012)
3. ASTM F623-99: Standard Performance Specification for Foley Catheter (2006)
4. T.R. Kucklick, The medical device R&D handbook (2013). http://www.crcnetbase.com/isbn/978-1-4398-1195-5%5Cnhttp://mirlyn.lib.umich.edu/Record/012219495. CN-R856.15.M43 2013
5. R.M. Donlan, Biofilms and device-associated infections. Emerg. Infect. Dis. **7**, 277–281 (2001). https://doi.org/10.3201/eid0702.700277
6. L.E. Nicolle, The chronic indwelling catheter and urinary infection in long-term-care facility residents. Infect. Control Hosp. Epidemiol. **22**, 316–321 (2001). https://doi.org/10.1086/50 1908
7. L. Muzzi-Bjornson, L. Macera, Preventing infection in elders with long-term indwelling urinary catheters. J. Am. Acad. Nurse Pract. **23**, 127–134 (2011). https://doi.org/10.1111/j.1745-759 9.2010.00588.x
8. C. Clec'h, C. Schwebel, A. Français, D. Toledano, J.-P. Fosse, M. Garrouste-Orgeas, E. Azoulay, C. Adrie, S. Jamali, A. Descorps-Declere, D. Nakache, J.-F. Timsit, Y. Cohen, Does catheter-associated urinary tract infection increase mortality in critically ill patients? Infect. Control Hosp. Epidemiol. **28**, 1367–1373 (2007). https://doi.org/10.1086/523279
9. M.M. Gabriel, *Effects of Silver on the Adherence of Escherichia coli and Other Bacteria to Urinary Catheters* (Georgia State University, Georgia, 1993)
10. B. Trautner, R. Darouiche, Role of biofilm in catheter-associated urinary tract infection. Am. J. Infect. Control **32**, 177–183 (2004). https://doi.org/10.1016/j.ajic.2003.08.005.Role
11. M. Bračič, *Surface Modification of Silicone with Polysaccharides for the Development of Antimicrobial Urethral Catheters* (Maribor, 2016)
12. S.M. Jacobsen, D.J. Stickler, H.L.T. Mobley, M.E. Shirtliff, Complicated catheter-associated urinary tract infections due to *Escherichia coli* and *Proteus mirabilis*. Clin. Microbiol. Rev. **21**, 26–59 (2008). https://doi.org/10.1128/CMR.00019-07
13. D. Choudhury, A. Thompson, V. Stojanoff, S. Langermann, J. Pinkner, S.J. Hultgren, S.D. Knight, X-ray structure of the FimC-FimH chaperone-adhesin complex from uropathogenic *Escherichia coli*. Science (80-.) **285**, 1061–1066 (1999). http://science.sciencemag.org/conte nt/285/5430/1061.abstract
14. D.R. Monteiro, L.F. Gorup, A.S. Takamiya, A.C. Ruvollo-Filho, E.R. de Camargo, D.B. Barbosa, The growing importance of materials that prevent microbial adhesion: antimicrobial effect of medical devices containing silver. Int. J. Antimicrob. Agents **34**, 103–110 (2009). https://d oi.org/10.1016/j.ijantimicag.2009.01.017
15. A. Mannan, S.J. Pawar, Anti-infective coating of gentamicin sulphate encapsulated PEG/PVA/chitosan for prevention of biofilm formation. Int. J. Pharm. Pharm. Sci. **6**, 571–576 (2014)

16. R.O. Darouiche, M.D. Mansouri, I.I. Raad, Efficacy of antimicrobial-impregnated silicone sections from penile implants in preventing device colonization in an animal model. Urology **59**, 303–307 (2002). https://doi.org/10.1016/S0090-4295(01)01533-3

17. B.W. Trautner, Management of Catheter-Associated Urinary Tract Infections (CAUTIs). Virology **23**, 76–82 (2010). https://doi.org/10.1097/QCO.0b013e328334dda8.Management

18. M. Bračič, L. Pérez, R. Martinez-Pardo, K. Kogej, S. Hribernik, O. Šauperl, L. Fras Zemljič, A novel synergistic formulation between a cationic surfactant from lysine and hyaluronic acid as an antimicrobial coating for advanced cellulose materials. Cellulose 1–17 (2014). https://doi.org/10.1007/s10570-014-0338-8

19. L. Fras Zemljič, J. Volmajer, T. Ristic, M. Bracic, O. Sauperl, T. Kreže, Antimicrobial and antioxidant functionalization of viscose fabric using chitosan–curcumin formulations. Text. Res. J. (2013). https://doi.org/10.1177/0040517513512396

Chapter 3
Polysaccharides in Medical Applications

Polysaccharides are the most important organic raw materials, building blocks in several fields and machineries of life [1]. Since the 1980s, naturally occurring polysaccharides, represented by cellulose, have been re-evaluated as outstanding chemicals and/or materials with various uses. Natural polysaccharides from different sources exhibit some special characteristics at the molecular and supramolecular levels, which are associated with their hydrogen-bonding ability, side-group reactivity, which can be modified covalently or by ionic bonds, enzymatic degradability, chirality, semi-rigidity, etc. Their higher-order structure involve their ability to form fibrous crystalline entities, chelate complexes, lyo-gels, liquid crystals, etc. [2–5].

Therefore, polysaccharides have found applications in clothing-, paper-, food-pharmaceutical industry, medical and construction fields, and many others. It is the medical field, however, where polysaccharides have made the greatest breakthrough in the past decade [6]. Most interest in polysaccharides was shown in the fields of hygiene products, regenerative medicine and nano-delivery systems [2–5]. Polysaccharides may be used for a functionalization of variety of medical devices, including catheters, vascular grafts, guidewires, and sensors. For example, the use of hyaluronic acids coatings can improve device biocompatibility and lubricity and reduce fouling and tissue abrasion [3, 5, 6]. In all mentioned fields, application of specific polysaccharides as biomaterials is of great importance since they provide good biocompatibility, low toxicity, and suitable biodegradability. Furthermore, specific polysaccharide structures provide interesting platforms for different functionalization and derivatisation in order to achieve water solubility, antimicrobial activity, antioxidant efficiency or antiviral properties [2–8]. There has also been an increasing demand in recent years for the application of natural products due to the increased health awareness and to address problems in the environment. Thus, the natural polysaccharides obtained from natural sources, such as plants or animals are of a great priority when applied for medical materials. Exemplary natural biodegradable polysaccharides include amylose, maltodextrin, amylopectin, starch, dextran, hyaluronic acid, heparin, chondroitin sulfate, dermatan sulfate, heparan sulfate, keratan sulfate, dextran sulfate, pentosan polysulfate, and chitosan [2–4]. Carboxymethyl cellulose and

M. Bračič et al., *Bioactive Functionalisation of Silicones with Polysaccharides*,
Biobased Polymers, https://doi.org/10.1007/978-3-030-02275-4_3

hyaluronic acid, anionic polysaccharides representatives, are known as moisturizing agents themselves or as an hydrophilic components or coatings integrated onto different solid matrices used for medical purposes [2, 3, 9, 10]. Different processing techniques and uses for hyaluronan have been invented and patented by Balazs, Leshchiner, and their coworkers [9]. In a very commonly used process, the hyaluronan was extracted from animal blood, deproteinised and washed with chloroform. This yielded a high-molecular weight hyluronan which was non-inflammatory, pyrogen-free, non-antigenic and was of high purity, which means that it was free of proteins and peptides as well as nucleic acid impurities. The product was bought and marketed by Pharmacia as Healon [9]. A large number of medical applications have already been described and successfully implemented for hyaluronan. Most notable being improvement of pathological joint function, improvement of eye functions, and prevention of postoperative adhesion tissues and tendons. More new application proposals are emerging every day as the use of natural materials in medical applications is growing largely in the last few decades. Especially in the field of medical implants, hyluronan has shown to have specific interactions with biomolecules and can be used as a coating material [2, 3, 10]. Beside these two anionic polysaccharides, alginates are very useful to moderate medical materials [11]. Medicinal properties of alginates, which are used to heal wounds and cure infections and other ailments, are driving the alginates market in most developing regions in a wide array of medical applications (dental impression moulds, prosthetics, wound and burn dressings, scaffolds and myocardial implantation; local delivery of gene therapy, etc.) [2, 3, 11]. Amino-functional polysaccharides, most promising as antimicrobial substances, are highly interesting for medical materials as they can be used as alternatives to conventional antimicrobial substances [12]. These polysaccharides contain amino groups which interact with the cell surface of pathogen microorganisms and in this way destroy them by several possible mechanisms [13]. One of the most popular amino polysaccharides is chitosan, which is obtained by alkaline deacetylation of chitin. Chitosan's positive charge, the degree of N-deacetylation, the mean polymerization degree and the nature of chemical modifications are the properties, which strongly influence its antimicrobial effectiveness [12, 13]. It is used as a natural antimicrobial agent for the new medical products development and is gaining on popularity as the Food and Drug Administration (FDA) approves it as a food ingredient. Besides its antimicrobial activity, it exhibits anti-cholester-olemic, anti-ulcer, anti-uremic and anti-tumour effects [14]. A wide variety of chitosan derivatives is synthesised due to the limited solubility of commercial chitosan. Amongst the most promising ones are quarternised chitosan, carboxymethylated chitosan and tiol–chitosan, which have found applications as matrixes for the preparation of several medical materials in different forms and for different purposes [15]. Some sulphated polysaccharides, like fucoidan, sulphated dextran, sulphated chitosan or sulphated cellulose have the potential to form surfaces with significant anticoagulant properties. Furthermore, some alternative sulphated polysaccharides like sulphated wood derived galactoglucomannans and xylans from different sources have been investigated regarding their antithrombotic properties and showed promising results [16–19]. The use of polysaccharides as building blocks in the development of nano-sized drug delivery systems is rapidly growing, thus the control over the size and shape of the particles has greatly improved in the last few years [3]. This can be attributed to the outstanding virtues of polysaccharides such as biocompatibility, biodegradability, low toxicity and low cost. In addition, the variety of physicochemical properties and the ease of chemical modifications enable the preparation of a wide array of nanoparticles. The use of carboxymethyl tamarind to coat silver nanoparticles has shown to improve the initial antimicrobial activity as the polysaccharide coating prevented the formation of a resistant bacterial biofilm [20]. Polysaccharide coatings also add to the biocompatibility of particles as they show low or no toxicity towards mammalian cells [20]. One can also prepare nanoparticles directly from polysaccharides, which find use as drug delivery systems. A simple ionic gelation of alginate with calcium chloride forms gelatinous particles which can entrap drugs and release them in a controlled fashion by changing the surrounding stimuli [21, 22]. The stability and targeting effect of such particles can be improved by addition of second polysaccharide like chitosan [22]. Recently several authors reported successful application of polysaccharide scaffolds in tissue engineering. A biomimetic and cytocompatible scaffold, manufactured from the combination of sodium alginate and silk fibroin showed the ability to promote cellular attachment and proliferation of skin cells [23]. Cellulose has also shown to be suitable for manufacturing of scaffolds for tissue regeneration. In a recently published work, bacterial cellulose from Acetobacter organism was grafted with positively charged moieties and exhibited high potency for cell attachment [24]. Even though natural-based and biodegradable materials offer a high level of biocompatibility and low toxicity, caution has to be taken before introducing them into the human body. One must understand that an initially non-toxic natural molecule or polymer can produce potentially harmful molecular species as it biodegrades in the body. Such species are often not found in the body or can be found in the body only at very low concentrations. Thus, the body can respond to these unknown species, causing unwanted side effects such as immune reactions, toxic build-up of the degradation products in the liver, or the initiation or provocation of other adverse effects on cells or tissue in the body [4]. Another drawback of natural-based biodegradable materials is their nature. Their chemical structure depends on the plant or animal that it was extracted from and the environmental conditions in which the plant or animal where grown. It is therefore difficult to ensure consistency in the chemical composition of natural-based materials [4].

Only a few representative approaches in the field of polysaccharide medical applications were represented above. It has been shown that some polysaccharides are very well suitable for manufacturing of different materials and obtaining special surface functionalities. The use of polysaccharides as coatings for silicone-based implants to achieve biofilm formation resistance is in more detailed described in the following chapter. It will be clearly shown that polysaccharides are becoming more and more popular as active catheter coatings.

3.1 Polysaccharide Coatings

Polysaccharide coatings for medical applications are still struggling to compete against other polymer or metal coatings for medical devices, which is mostly due to their underperformance in many cases. Nevertheless, the vast research efforts in this field have greatly improved the quality of such coatings and their value continues to grow [25–27]. The use of natural biodegradable polysaccharides provides many advantages when applied to the surface of different medical devices/materials, thus the following text and the next few sections are dedicated to various methods/technologies, which have been developed for incorporation of polysaccharides onto materials depending on the particular active agent and basic material type. Generally speaking, polysaccharides can be incorporated into the polymer solution before material processing (prior to the extrusion of the polymer mass or blended into the fibres during their formation) [4] or can be applied onto produced materials as surface coatings. Regarding the latest, surface coatings provide better efficiency. Polysaccharides alone or in combination with other active agents can be applied to the surface of a medical device in different ways. There is not a single, universally accepted classification of coating technologies, but according to Romanó et al., one can divide them in three groups according to their strategy of action [28]:

1. Passive coatings. These are often referred to as passive surface modification (PSM), which refers to surface coatings that prevent or reduce bacterial adhesion to the implant, but do not release any antibacterial substances from the surface to the surrounding tissue.
2. Active coating. These are often referred to as active surface modification (ASM), which refers to surface coatings that have antimicrobial substances like antibiotics, metal ions and others incorporated in their structure. These substances are released from the coating in order to kill the bacteria in the surrounding tissue. Polysaccharides are emerging as active coatings recently as they are widely accepted as good drug carriers.
3. Peri-operative coatings. These are often referred to as local carriers or coatings (LCC). LCC's are basically antibacterial coatings that are applied to medical devices or implants in situ during surgery. They are meant to protect the site of surgery from bacteria during implant insertion as well as after implantation [28].

Polysaccharide coatings are nowadays used in various medical fields, such as, nanotechnology, wound healing and tissue regeneration, gene therapy, etc. [5, 11, 19]. For example, hyaluronic acid is a very common representative of a widely used polysaccharide in medicine and is used as a coating of polyurethanes, polyesters, polyolefins and many more. A lot of these coatings are patented, such as the hyaluronic acid crosslinked gels with incorporated active substances including, other types of hydrophilic polysaccharides, proteins, and synthetic water soluble polymers [5, 29, 30]. The high hydrophilicity of hyaluronic acid, which is due to its high water adsorption capacity, makes it very suitable as a coating for medical devices with the aim to repel bacteria and other biomolecules, which generally prefer hydrophobic surfaces. The high adsorption capacity is also utilised to incorporate antibiotics in the hyaluronic acid gels and release them in a controlled fashion. Such coatings have shown to reduce the adhesion of *Staphylococcus epidermis* and are used to coat polyurethane catheters and silicone shunts [30]. Another very common polysaccharide widely used as a coating in the medical field is heparin. Its unmatched anti-coagulative properties are exploited in the coating world to reduce fibronectin deposition on vascular implants, which prevents thrombosis and colonisation of bacteria. In a recent clinical trial, 246 patients were subjected to cardiovascular catheter insertion. The catheters were either uncoated or coated with heparin. The trial showed that the CAUTI were significantly reduced due to the inability of bacteria to colonise the catheters [30]. Indest et al. [19] applied the selected polysaccharides (chitosan, fucoidan, and chitosan sulfate) on the polyethylene terephthalate (PET) surfaces as potential coating for developing of antimicrobial or anticoagulant PET surfaces. The adsorption on PET model films was monitored using a quartz crystal microbalance with a dissipation unit. The surface chemistry and morphology of the chitosan/fucoidan or chitosan/chitosan sulphate coated PET-H films was analyzed using X-ray photoelectron microscopy (XPS) and atomic force microscopy (AFM). It was found out that chitosan/fucoidan films were thinner and more compressed, while in the case of chitosan/chitosan sulphate, large amounts of chitosan sulphate were adsorbed, indicating a loose and thick adsorbed film. Furthermore, the influence of different modifications of PET surfaces with above selected polysaccharides (chitosan, fucoidan, chitosan sulphate) on human serum albumin (HSA) adsorption was evaluated [19]. Protein repellent properties indicate the anticoagulant and antifouling surface properties. It was found that due to steric repulsion the chitosan, fucoidan and chitosan sulphate layers reduced HSA adsorption, especially in case of chitosan/fucoidan and chitosan/chitosan sulphate covered films. The coatings showed high potential for modification of vascular grafts modifications to obtain anticoagulant properties [19, 28]. It has also been shown that sulphated xylans from different sources (hard wood and oat-spelt) possess antithrombotic properties and can be as coatings applied onto synthetic vascular grafts [31]. Two types of xylans, glucuronoxylan derived from beech wood, and arabinoxylan from oat spelt, were sulphated. The results showed significant increases in negative charges for the sulphated samples, which were a consequence of introducing sulphated groups as strong acids. The increase of antithrombotic properties was not influenced only by the presence of certain amounts of sulphate functional groups but also by the total negative-charges originating from both sulphate and carboxyl groups. The later was proved by the high linear correlation between the activated partial thromboplastin time values, and the total charge of the samples.protonation/deprotonation behavior in water solutions influence these properties [31]. Low-density polyethylene is widely used medical devices and therefore it shares the major problem of bacterial infections upon introduction in the body with other implantable materials [32]. To resolve this problem, the authors have coated polyethylene films with a natural alginic acid by covalent binding using oxygen plasma to activate the originally inert polyethylene films. The results have shown that the net charge, hydrophilicity and the amount of antimicrobial agent significantly influence the overall antibacterial performance. It has been demonstrated that the polyanionic alginic acid specifically triggers the adherence of the cytoplas-mic membrane of *Escherichia coli* bacteria, which is made of lipopolysaccharides, opposed to the *Staphylococcus aureus* membrane made of peptidoglycans. This find-ing showed that polysaccharides can act not only as antimicrobial agents, but also have some specific actions against different types of bacterial cell membranes [32].

More recently, biomacromolecules-coatings have been incorporated into orthopaedic implants in order to modulate the surrounding biological environment [33]. Some polysaccharides like hyaluronic acid and chitosan possess the ability to prevent bacterial adhesion and/or bactericidal proliferation and activity. It was reported that osteoblast adhesion is impaired by the presence of the hyaluronic acid chains. The influence of different pH values on chitosan macromolecules' confor-mation and consequently on its adsorption ability and coating morphology, as well as on antimicrobial activity of coated fibres used for medical purposes was studies by Ristič et al. [34]. Chitosan coated materials exhibit higher overall inhibition of microorganisms than trimethyl chitosan-coated materials. Microbiological testing confirmed that fibres treated with trimethyl chitosan are ineffective against *E. coli*. They also show a reduced activity against *C. Albicans*. This suggests that chitosan is a suitable and effective antimicrobial agent for fibre coating and the development of antimicrobial medical textiles. Additionally, a model drug was incorporated in the chitosan nanoparticles which were attached to fibres in order to create potential fibrous drug delivery systems. The developed fibrous drug-loaded system demon-strated the potential for gynaecological curative treatment. Besides effective drug release and antimicrobial efficiency, the design of tampons may be adjusted to cervi-cal anatomy (release directly into the cervical mucosa) [26]. Low-density polyethy-lene films were coated by alginic acid by Karbassi et al. [32] in order to improve antibacterial properties. The authors have coated polyethylene films with a natural polysaccharide alginic acid by covalent binding using oxygen plasma to activate the originally inert polyethylene films. Allylalcohol (AAL), allylamine (AAM), and 2-hydroxyethyl methacrylate (HEMA), respectively, were attached to the film to form polymeric brushes (Fig. 3.1). The alginic acid (ALGA) was later bound to such modified polyethylene films by carbodiimide chemistry [32].

The antibacterial tests were performed against the Gram negative *Escherichia coli* and the Gram positive *Staphylococcus aureus*. The results have shown that the net charge, hydrophilicity and the amount of antimicrobial agent significantly influ-ence the overall antibacterial performance. This has been directly linked to the type of grafting used to attach the antimicrobial agent to the polyethylene surface. The bonding strength and the surface chemistry influence attachment of the bacteria to the surface and their later interactions, which lead to cell membrane destruction. It has been demonstrated that the polyanionic alginic acid specifically triggers the adher-ence of the cytoplasmic membrane of *Escherichia coli* bacteria, which is made of lipopolysaccharides, opposed to the *Staphylococcus aureus* membrane made of pep-tidoglycans. This finding not only show that polysaccharides can act as antimicrobial agents but also have some specific actions against different types of bacterial cell membranes [32]. In order to achieve corrosion resistance, biocompatibility, improved ostseointegration, as well as prevention of inflammation and pain, Finšgar et al. [25]

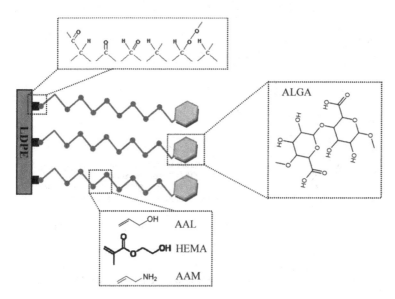

Fig. 3.1 A schematic representation of low-density polyethylene film coating with alginic acid via carbodiimide binding using polymer brushes as anchoring points [32]. Reprinted from Karbassi et al. [32], Copyright (2014), with permission from the authors; licensee MDPI, Basel, Switzerland

used cationic polysaccharide chitosan to coat steel hip implants. By incorporation of the non-steroid anti-inflammatory drug, diclofenac, into the multi-layered coatings the prevention of inflammation and pain was achieved. The multilayer coatings were prepared on steel plates by spin-coating an aqueous chitosan solution. A thin chitosan coating is formed after this step, which is followed by spin-coating of an aqueous diclofenac solution. This process was repeated several times to form multilayers. The corrosion resistivity was reduced if the final layer was made of diclofenac, indicating the importance of the polysaccharide against general corrosion of the steel plate. The drug release tests, depicted in Fig. 3.2, have shown that a burst release of diclofenac can be observed in the first hour where 50% of the drug is released. In the next five hours, the drug was released in a controlled and rather fast fashion where over 90% of the drug was released. The rest was released in a slow fashion over the course of one day [25].

Two tests were performed in order to evaluate the biocompatibility of the multilayer chitosan-diclofenac coating on steel plates. In one case the solutions of the drug release test after certain time intervals were subjected to human osteoblast cells and their viability was measured. In the second case the viability of the osteoblasts was measured directly on the coated steel surface and a Live/Dead assay was used for this purpose. The confocal scanning microscope images in Fig. 3.3 show that cell viability is high for the coated steel surface but extremely low for the uncoated surface, indicating that the polysaccharide coating promotes cell proliferation [25].

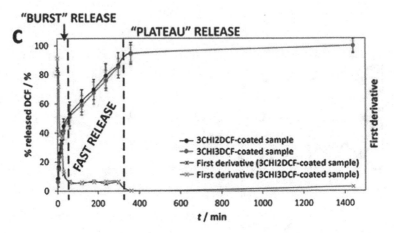

Fig. 3.2 Percentage of diclofenac released from the chitosan multilayer coating over time [25]. Reprinted from Finšgar et al. [25], Copyright (2016), with permission from Macmillan Publishers Limited

Fig. 3.3 Confocal scanning microscopy images of the Live/Dead assay of human osteoblast cells on a neat steel surface and on steel surfaces coated with the chitosan-diclofenac multilayers [25]. Reprinted from Finšgar et al. [25], Copyright (2016), with permission from Macmillan Publishers Limited

Only a few representative approaches in the field of polysaccharide medical applications were represented above. It was shown that some polysaccharides are very well suitable for surface coating of different materials and obtaining special surface functionalities. The use of polysaccharides as coatings for silicone-based implants in order to achieve biofilm formation resistance is in more detailed described in the following chapter.

References

1. P. Berlin, D. Klemm, J. Tiller, R. Rieseler, A novel soluble aminocellulose derivative type: its transparent film-forming properties and its efficient coupling with enzyme proteins for biosensors. Macromol. Chem. Phys. **201**, 2070–2082 (2000). https://doi.org/10.1002/1521-39 35(20001001)201:15%3c2070:AID-MACP2070%3e3.0.CO;2-E
2. S. Dumitriu (ed.), *Polysaccharides in Medicinal Applications* (Marcel Dekker Inc., New York, 1996)
3. V. Popa (ed.), *Polysaccharides in Medicinal and Pharmaceutical Applications* (Smithers Rapra Technology, 2011)
4. S. Chudzik, *Natural Biodegradable Polysaccharide Coatings for Medical Articles* (W.I.P. Organization, 2005)
5. A. Basu, K.R. Kunduru, E. Abtew, A.J. Domb, Polysaccharide-based conjugates for biomedical applications. Bioconjug. Chem. **26**, 1396–1412 (2015). https://doi.org/10.1021/acs.bioconjch em.5b00242
6. P. Mandal, C.A. Pujol, E.B. Damonte, T. Ghosh, B. Ray, Xylans from *Scinaia hatei*: structural features, sulfation and anti-HSV activity. Int. J. Biol. Macromol. **46**, 173–178 (2010). https://doi.org/10.1016/j.ijbiomac.2009.12.003
7. H. Fasl, J. Stana, D. Stropnik, S. Strnad, K. Stana-Kleinschek, V. Ribitsch, Improvement of the hemocompatibility of PET surfaces using different sulphated polysaccharides as coating materials. Biomacromolecules **11**, 377–381 (2010). https://doi.org/10.1021/bm9010084
8. A. Salam, J.J. Pawlak, R.A. Venditti, K. El-tahlawy, Incorporation of carboxyl groups into xylan for improved absorbency. Cellulose **18**, 1033–1041 (2011). https://doi.org/10.1007/s10 570-011-9542-y
9. Y. Nishio, *Material Functionalization of Cellulose and Related Polysaccharides via Diverse Microcompositions* (Springer, Berlin, 2006)
10. G. Kogan, L. Šoltés, R. Stern, P. Gemeiner, Hyaluronic acid: a natural biopolymer with a broad range of biomedical and industrial applications. Biotechnol. Lett. **29**, 17–25 (2007). https://do i.org/10.1007/s10529-006-9219-z
11. M.J. Franklin, D.E. Nivens, J.T. Weadge, P.L. Howell, Biosynthesis of the *Pseudomonas aeruginosa* extracellular polysaccharides, alginate, Pel, and Psl. Front. Microbiol. **2**, 167 (2011). https://doi.org/10.3389/fmicb.2011.00167
12. S.U. Shen (ed.), *Pullulan Films and Their Use in Edible Packaging* (W.I.P. Organization, 2007)
13. M.N.V. Ravi Kumar, A review of chitin and chitosan applications. React. Funct. Polym. **46**, 1–27 (2000). https://doi.org/10.1016/S1381-5148(00)00038-9
14. R. Ahvenainen, Active and intelligent packaging: an introduction, in *Novel Food Packaging Techniques* (Woodhead Publishing Limited, 2003)
15. E. Shalaby (ed.), *Biological Activities and Application of Marine Polysaccharides* (InTech Open, 2017)
16. A. Doliška, S. Strnad, J. Stana, E. Martinelli, V. Ribitsch, K. Stana-Kleinschek, In vitro haemocompatibility evaluation of PET surfaces using the quartz crystal microbalance technique. J. Biomater. Sci. Polym. Ed. **23**, 697–714 (2012). https://doi.org/10.1163/092050611X559232
17. M. Gericke, A. Doliška, J. Stana, T. Liebert, T. Heinze, K. Stana-Kleinschek, Semi-synthetic polysaccharide sulfates as anticoagulant coatings for PET, 1 - cellulose sulfate. Macromol. Biosci. **11**, 549–556 (2011). https://doi.org/10.1002/mabi.201000419
18. D. Stephan, P. Katrin, K. Manuela, B. Anja, S.U. Suhubert, H. Thomas, Homogeneous sulfation of xylan from different sources. Macromol. Mater. Eng. **296**, 551–561 (2011). https://doi.org/10.1002/mame.201000390
19. T. Indest, J. Laine, L.-S. Johansson, K. Stana-Kleinschek, S. Strnad, R. Dworczak, V. Ribitsch, Adsorption of fucoidan and chitosan sulfate on chitosan modified PET films monitored by QCM-D. Biomacromol **10**, 630–637 (2009). https://doi.org/10.1021/bm801361f
20. S. Sanyasi, R.K. Majhi, S. Kumar, M. Mishra, A. Ghosh, M. Suar, P.V. Satyam, H. Mohapatra, C. Goswami, L. Goswami, Polysaccharide-capped silver Nanoparticles inhibit biofilm formation

and eliminate multi-drug-resistant bacteria by disrupting bacterial cytoskeleton with reduced cytotoxicity towards mammalian cells. Sci. Rep. **6**, 1–16 (2016). https://doi.org/10.1038/srep 24929

21. F. Sarei, N.M. Dounighi, H. Zolfagharian, P. Khaki, S.M. Bidhendi, Alginate nanoparticles as a promising adjuvant and vaccine delivery system. Indian J. Pharm. Sci. **75**, 442–449 (2013). https://doi.org/10.4103/0250-474X.119829

22. F. Wang, S. Yang, J. Yuan, Q. Gao, C. Huang, Effective method of chitosan-coated alginate nanoparticles for target drug delivery applications. J. Biomater. Appl. **31**, 3–12 (2016). https://doi.org/10.1177/0885328216648478

23. Y. Wang, X. Wang, J. Shi, R. Zhu, J. Zhang, Z. Zhang, D. Ma, Y. Hou, F. Lin, J. Yang, M. Mizuno, A biomimetic silk fibroin/sodium alginate composite scaffold for soft tissue engineering. Sci. Rep. **6**, 1–13 (2016). https://doi.org/10.1038/srep39477

24. J.C. Courtenay, M.A. Johns, F. Galembeck, C. Deneke, E.M. Lanzoni, C.A. Costa, J.L. Scott, R.I. Sharma, Surface modified cellulose scaffolds for tissue engineering. Cellulose **24**, 253–267 (2017). https://doi.org/10.1007/s10570-016-1111-y

25. M. Finšgar, A.P. Uzunalić, J. Stergar, L. Gradišnik, U. Maver, Novel chitosan/diclofenac coatings on medical grade stainless steel for hip replacement applications. Sci. Rep. **6**, 1–17 (2016). https://doi.org/10.1038/srep26653

26. T. Ristić, A. Zabret, L.F. Zemljič, Z. Peršin, Chitosan nanoparticles as a potential drug delivery system attached to viscose cellulose fibers. Cellulose **24**, 739–753 (2017). https://doi.org/10.1 007/s10570-016-1125-5

27. M. Bračič, L. Fras-Zemljič, L. Pérez, K. Kogej, K. Stana-Kleinschek, R. Kargl, T. Mohan, Protein-repellent and antimicrobial nanoparticle coatings from hyaluronic acid and a lysine-derived biocompatible surfactant. J. Mater. Chem. B. **5**, 3888–3897 (2017). https://doi.org/10. 1039/C7TB00311K

28. C.L. Romanò, S. Scarponi, E. Gallazzi, D. Romanò, L. Drago, Antibacterial coating of implants in orthopaedics and trauma: a classification proposal in an evolving panorama. J. Orthop. Surg. Res. **10**, 157 (2015). https://doi.org/10.1186/s13018-015-0294-5

29. G. Scott (ed.), *Degradable Polymers: Principles and Applications*, 2nd edn. (Springer, Netherlands, 2002)

30. F. Iolanda, D. Gianfranco, Prevention and control of biofilm-based medical-device-related infections. FEMS Immunol. Med. Microbiol. **59**, 227–238 (2010). https://doi.org/10.1111/j.1 574-695X.2010.00665.x

31. S. Strnad, N. Velkova, B. Saake, A. Doliška, M. Bračič, L.F. Zemljič, Influence of sulfated arabino- and glucuronoxylans charging-behavior regarding antithrombotic properties. React. Funct. Polym. **73**, 1639–1645 (2013). https://doi.org/10.1016/j.reactfunctpolym.2013.09.007

32. E. Karbassi, A. Asadinezhad, M. Lehocký, P. Humpolíček, A. Vesel, I. Novák, P. Sáha, Antibacterial performance of alginic acid coating on polyethylene film. Int. J. Mol. Sci. **15**, 14684–14696 (2014). https://doi.org/10.3390/ijms150814684

33. S.B. Goodman, Z. Yao, M. Keeney, F. Yang, The future of biologic coatings for orthopaedic implants. Biomaterials **34**, 3174–3183 (2013). https://doi.org/10.1016/j.biomaterials.2013.0 1.074

34. T. Ristić, S. Hribernik, L. Fras-Zemljič, Electrokinetic properties of fibres functionalised by chitosan and chitosan nanoparticles. Cellulose **22**, 3811–3823 (2015). https://doi.org/10.1007/ s10570-015-0760-6

Chapter 4
Functionalisation of Silicones with Polysaccharides

A basic strategies to deal with the above described problems of CAUTI and biofilm formation is aimed toward modification of the implant's surface-chemical properties, coating with a desired agent, and by manipulation of the surface roughness or morphology which can prevent the attachment of bacteria to the implant [1]. This chapter will be devoted predominantly to coating of silicon-based medical implants. Two strategies are common in coating of medical implants to achieve the above-mentioned goals. One is coating with an antimicrobial agent which allows a slow and controlled release from the surface with time, and the other is a permanent non-migrating coating which prevents the attachment of microorganisms and kills them at the implants surface (Fig. 4.1) [1].

In Fig. 4.1 left, the antimicrobial substance is incorporated in a polymer matrix which allows the slow and controlled release of the substance. In Fig. 4.1 right, the

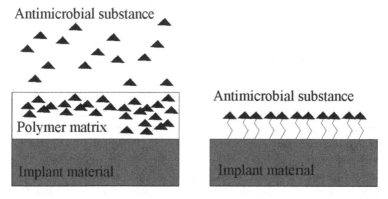

Fig. 4.1 Common strategies for coating of medical implants. Incorporation of an antimicrobial substance in a polymer matrix attached to the implant's surface for controlled release (left) and permanent attachment of a non-migrating antimicrobial substance directly to the implant's surface (right) [2]

M. Bračič et al., *Bioactive Functionalisation of Silicones with Polysaccharides*, Biobased Polymers, https://doi.org/10.1007/978-3-030-02275-4_4

27

polymer bound to the implants surface itself has antimicrobial functionalities. No additional polymer is needed for its binding, but also no migration of the antimicrobial functionalities is possible. Both approaches have their advantages and disadvantages. The leaching of the antimicrobial substance from the implants surface has the advantage of killing the microorganisms in the surrounding environment of the implant, which allows a slow and controlled release, but its effect decreases over time. On the other hand, the permanently bound antimicrobial substance keeps its effectiveness over a longer period, but has no effect on microorganisms in the surrounding environment. It is active only in the vicinity to the surface i.e. interface between the surface of the implant and the surroundings. It is also worth mentioning that the second approach is favourable when seeking for regulatory approval of the medical implant, because migrating substances must be tested for their influence on the body as a whole and not only for the part they were used on [1]. It must nevertheless be considered that there is no universal strategy that can be used for all types of implants and therefore one must tackle the problem by considering the use and placement in the body of the specific implant.

The silicone surfaces coated in this work are meant to be used as urethral catheters, which are long tubes with a rather small lumen diameter. The free-flowing bacteria in the urine during a CAUTI preferably adsorb to the catheter surface and form a biofilm which can clog the catheter in a short period of time [3, 4]. Therefore, it is necessary for the coating to prevent or minimize the biofilm formation, by disrupting the attachment process and by killing the adherent bacteria [5, 6]. To approach this issue one must choose the appropriate antimicrobial or antifouling substance and have the knowledge of its physicochemical properties and the physicochemical properties the material the implant/urethral catheter is made of [3, 7]. As silicones are inert materials with a low surface free energy, pre-treatment (activation) of their surface is often required in order to increase their chemical reactivity allowing for improved silicone-bioactive agent interactions.

4.1 Activation of Silicone Surfaces

It is well established that the low surface free energy silicones are notoriously difficult to be modified by surface interactions with other materials. This lead to the application of various chemical and physical surface altering methods to activate the surface of silicones and increase their surface free energy [8–10]. It is due to the rather inert $-CH_3$ moieties on the backbone of the $-Si-O$ chain of the silicones that chemical interaction with other chemical species is difficult. Altering this by high-energy etching methods or chemical oxidation yields a silicone surface comprised of various new functional groups that are highly reactive and increase the surface free energy of the silicone (e.g. hydroxyl groups, carboxyl groups, free oxygen radicals, amino groups, etc.) [8, 11, 12]. These functionalities also convert the initially hydrophobic silicone surface to a hydrophilic one allowing for a better wetting of the surface when binding of other molecules is performed in polar environment. Hydrophilicity of a silicone surface is especially beneficial in the case of urethral catheters as it also allows for an irritation-free insertion of the catheter, minimizing the discomfort of patients [3, 7, 13]. Hydrophilic surfaces are also known to reduce the adhesion of biomolecules on the surface which can contribute greatly to the reduction of biofilm formation [14, 15]. A general approach to activate, modify and increase the reactivity of silicone surfaces is to modify its surface chemical properties. This is often done by methods based on ionized ions interacting with the surface atoms of silicone like plasma [9, 10, 12] or UV/Ozone [8, 16–18] etching and corona discharge [11]. Another option of chemical alteration is hydrolysis of the silicone surface by means of strong acids [19, 20]. In both cases however, immediate bonding of the free radicals formed on the silicone surface with other molecules is needed as the surface reduces its high surface energy by reorientation of atoms and diffusion of low molecular weight fragments in the bulk, thus largely recovering its original inert character [8, 18]. The activation and hydrophobic recovery studies of PDMS by UV/ozone performed by Efimenko et al. [8] revealed that the PDMS macromolecules in the surface region undergo chain scission, involving both the main chain backbone and the side groups. In contrast, the UV/ozone activation causes significant changes in the surface structure of PDMS. The etching by molecular oxygen and ozone created during the UVO process results in a surface containing a large number of hydrophilic (mainly –OH) groups. The advancing water contact angle measurements in Fig. 4.2 showed that the UV/Ozone treatment increases the hydrophilicity of the PDMS surface and its intensity highly depends on the spectral intensity of the UV lamps and the ozone created in the process. Higher UV lamp intensity (UVO90: 90% radiation at 184.9 nm) introduces hydrophilic moieties to the PDMS surface faster than the lower intensity (UVO60: 65% radiation at 184.9 nm), which is reflected in lower contact angles at specific time points. The final contact angle value at 90 min is on the other hand the same for both lamp intensities used in this work. The ozone that is formed during the UV/ozone process shows a significant influence on the hydrophilicity of the PDMS surface. The contact angles reach a value of below 30° at 20 min exposure time while the UV treated sample, at the same time of exposure, exhibits a contact angle of over 90°. The UV treatment proved to be completely inefficient as the contact angles never dropped under 80°, underlining the vital influence of the ozone on the activation of the PDMS surface [8].

Fig. 4.2 Advancing water contact angles of PDMS surfaces activated with UV and UV/Ozone at different UV lamp intensities [8]. Reprinted from Efimenko et al. [8]. Copyright (2002), with permission from Elsevier

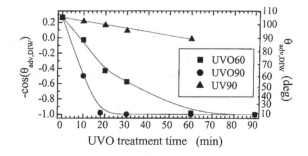

Similar effects of PDMS surface functionalisation can be achieved by means of plasma discharge. Plasma can be described as a (partially) ionized gas and is generated by applying either high temperatures or strong magnetic or electromagnetic fields to a gas. The latter method is used for polymer functionalization. Cold plasma treatment is an extremely versatile technique for modifying polymer surfaces. It has been reported that plasma treatment can improve polymer-polymer adhesion, the best results being obtained when using oxygen plasma. After the activation with plasma, the PDMS surface changed chemically and structurally. The result of surface activation using oxygen plasma is the formation of different oxygen containing polar functional groups such as C–O, C=O, and O–C=O, which act as nucleophilic centers to which adsorbent atoms can bind [21]. In an unpublished work of the authors of this book, PDMS became highly hydrophilic after oxygen plasma treatment due to the creation of new polar groups on the silicone surface, as well as the greater roughness of the material. The oxygen plasma used in this work was generated by radio-frequency with a power output of 500 W. The PDMS samples were treated by oxygen plasma for 3 and 10 s. As can be seen from Fig. 4.3, the untreated PDMS exhibited a uniform and rather smooth surface with an average surface roughness of 0.3 nm and a static water contact angle of 113°. When exposed to the oxygen plasma for 10 s at 500 W the surface gets clearly damaged. One can observe large evenly distributed spikes on the PDMS surface, which were attributed to the etching of the surface and the formation of SiO_x domains. The surface roughness of this surface was 6.6 nm and the static water contact angle was below 10° and difficult to accurately measure as the water drop rapidly spread over the whole surface. A high hydrophilicity and thus a high surface energy of the silicone without the physical damage of the surface was achieved by exposing the PDMS to oxygen plasma for 3 s. As can be seen in Fig. 4.3, the 3 s plasma sample exhibited a very smooth surface as the etching effect was very uniform and mild. The average surface roughness of this sample was 0.3 nm, making it slightly smoother than the pristine PDMS silicone surface. The static water contact angle of this sample was 15°.

Roth et al. [9] activated the silicone surface by oxygen and ammonia plasma. The PDMS was irradiated at different plasma power and different exposure time. It was shown that increasing the plasma power from 100 to 600 W can reduce the water contact angle from 38.4° to 26.3° at an exposure time of 15 s. The time of exposure had an even more significant influence as the contact angles reached values of 0° after the PDMS was exposed to a plasma of 600 W for 60 s. The ammonium plasma treatment rendered the PDMS surface hydrophilic as well. The plasma power in this case was only 10 W and the contact angle at this intensity reached a value of 62.2° after being exposed for 300 s. More important, using different plasma ions introduced different functional groups to the surface of the PDMS allowing for a variety of chemical bond options between the activated PDMS and other molecules. Binding other molecules (e.g. polymers) is necessary as hydrophobic recovery of the activated silicone surface is almost instant. The hydrophobic recovery of silicones is a known phenomenon. Fritz and Owen [22] described the hydrophobic recovery of plasma activated silicone two decades ago. They found out that a thin, wettable silica-like layer is produced with various plasma gases such as argon, helium, oxygen and nitrogen. However,

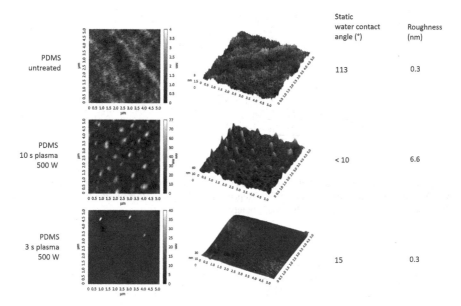

Fig. 4.3 AFM images (5 × 5 μm) of untreated and oxygen plasma treated PDMS silicone and their corresponding static water contact angles and surface roughness

in all cases the surface gradually recovered its initial hydrophobicity. Full recovery was in some cases achieved in 24 h while all samples where recovered completely after a week. They concluded that cracks forming on the surface of silicone due to plasma etching can be accounted for a rapid surface diffusion of low molecular weight material out of fresh cracks followed by slower bulk diffusion through the polymer matrix. Roth et al. [9] observed the same phenomenon in their work. The oxygen plasma activated silicone surface recovered almost completely after 360 min (Fig. 4.4). After 15 s in oxygen plasma (600 W) the surface of silicone had a contact angle of around 25°. It linearly increased with storage time, reaching a value of almost 80° after 360 min [9].

In order to overcome difficulties in surface modification processes of activated silicones due to hydrophobic recovery one should bind other molecules like polymers to the surface of the activated silicone immediately after activation. In Roth's work [9] this was overcome by grafting of poly(ethylene-*alt*-maleic anhydride) (PEMA) on the surface of the oxygen and ammonium plasma activated silicone (Fig. 4.5). Such modified silicone surfaces retained their hydrophilicity even after one week of storage in air.

Hemmilä et al. [10] performed a study of thirty nine different surface treatments to achieve long-term hydrophilicity of silicones. The wetabillity of the surfaces was observed over a period of six months. The surfaces were firstly activated by oxygen plasma and secondly coated with biocompatible polymers (polyvinylpyrrolidone, polyethylene glycol—PEG, 2-hydroxyethyl methacrylate) by physisorption or syn-

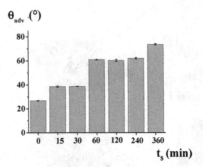

Fig. 4.4 Advancing water contact angles of a PDMS surface activated with oxygen plasma (15 s, 600 W) after storage in air for 360 min [9]. Reprinted (adapted) with permission from Roth et al. [9]. Copyright (2008) American Chemical Society

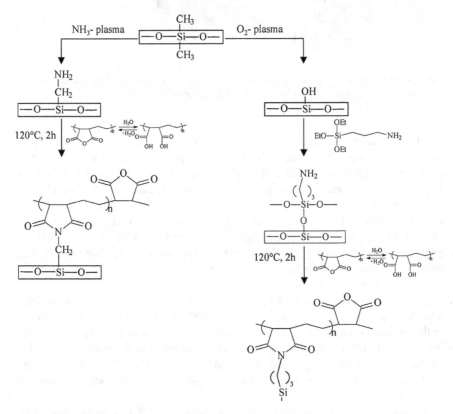

Fig. 4.5 Schematic representation of poly(ethylene-*alt*-maleic anhydride) grafting on oxygen and ammonium plasma activated PDMS [9]. Reprinted (adapted) with permission from Roth et al. [9]. Copyright (2008) American Chemical Society

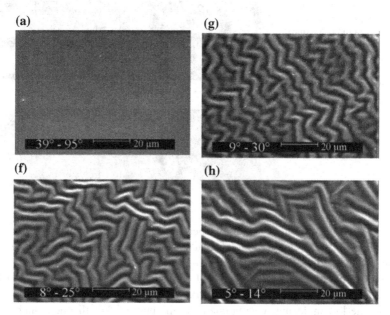

Fig. 4.6 Scanning electron microscopy (SEM) images of **a** neat PDMS, **f** 1 min PEG "grafted to" **g** 10 h PEG "grafted to", and **h** 10 h PEG "grafted from". Initial and long-term contact angles are shown in the image as well [10]. Reprinted from Hemmilä et al. [10]. Copyright (2012), with permission from Elsevier

thesis of both "grafting to" and "grafting from" polymer brushes. The grafted surfaces exhibited a well-defined nanostructure (Fig. 4.6) which provided the PDMS surface with long-term hydrophilicity even after 6 months of storage in air.

Using hydrophilic polysaccharides to achieve a permanent hydrophilicity of silicones was studied as well. Bracic et al. [23] used a patterned surface modification of PDMS silicone with cellulose and achieved a spatial alternating hydrophilic and hydrophobic areas on the surface of silicone (Fig. 4.7). Thin-films of PDMS silicone were manufactured on glass and gold substrates by spin-coating from a PDMS/Toluen solution. A square patterned metal mask was used to cover the thin films and activate the exposed silicone surface by UV/Ozone. A trimethylsilyl cellulose layer was deposited on the activated silicone surface by spray-coating or spin-coating and then converted to cellulose by a process named regeneration. The silyl groups are cleaved of by hydrochloric acid vapour and hydroxyl groups are formed, yielding cellulose.

Fig. 4.7 Fully (bottom) and spatially (top) cellulose-structured silicone surfaces stained with tetramethylrhodamine [23]. Reprinted from Bracic et al. [23]. Copyright (2014), with permission from Royal Society of Chemistry

Such modified silicone surfaces exhibit a permanent hydrophilicity with a water contact angle of 32°. The initial UV/Ozone activated silicon surface exhibited a water contact angle of 25° [23].

The hydrophobic recovery effect can also be suppressed by thermal aging of silicone during the curing process. Eddington et al. [24] showed that thermal aging removes the low-molecular weight chains in the bulk of the silicone, which were attributed to cause hydrophobic recovery by diffusing to the surface of the thermodynamically unstable silicone surface after oxygen plasma activation. The PDMS silicone samples were cured on a hotplate at 85 °C for 100 min and then thermally aged in an oven at 100 °C for 2–14 days. Water contact angle measurements revealed that unaged samples completely recovered their hydrophobicity (contact angle of over 95°) in 1–2 days while the sample thermally aged for 14 days did not recover completely, only reaching contact angle values of below 60°. The initial contact angle values after exposure to oxygen plasma were below 10° for all samples [24].

Besides etching techniques, chemical-based activation of silicones was reported as well. A subsequent "piranha" solution and potassium hydroxide chemical modification was reported by Maji et al. [20]. The PDMS silicone surface was firstly treated with a "piranha" solution consisting of hydrogen peroxide and sulphuric acid at a ratio of 2:3. Secondly, the silicone was dipped in a 1 M potassium hydroxide solution. In both cases the silicone was treated for 15 min. The results showed that a hydrophilic silicone surface was obtained with a water contact angle of 27°. The hydrophilicity was ascribed to be due to an increase of hydroxyl groups after the chemical treatment of the silicone [20].

An important factor of all mentioned activation techniques are the mechanical properties of the treated silicones and their surface morphology. Oláh et al. [18] activated PDMS by means of UV/ozone and fund out that silicones containing a homogenously dispersed filler exhibited a decrease in surface roughness, measured by AFM, when the oxidized surface rearranged into a smooth SiO_x layer with a surface roughness of below 2 nm. Silicones containing a heterogeneously distributed filler on the other hand, exhibited an increasing surface roughness to above 140 nm, which was attributed to the rearrangement of the oxidized surface, thus exposing the underlying filler aggregates [18]. Morphological damage in form of cracks was also observed by Haji et al. when activating silicone rubber with corona discharge [11]. Béhafy et al. [25] have investigated the correlation between the etching intensity of oxygen plasma and the elastic modulus of the newly formed silica-like layer as well as the crack formation on the surface of the PDMS. It was discovered that increasing the power of the plasma (5–60 W) at constant treatment time (60 s) or increasing the treatment time (10–300 s) at constant plasma power (20 W) equally decreased the elastic modulus of the top silica-like surface of the activated silicone. The initial elastic modulus for the experiment at 60 s treatment time at 5 W oxygen plasma was 7.1 GPa. It was reduced to 0.4 with increasing plasma power to 60 W. This sample also exhibited visible cracking of the top silica-like layer and formed buckles after bending tests, which showed that the silicone surface was permanently damaged. This was not the case for samples activated at lower plasma powers. Interestingly the effect of the plasma power and surface damage also reflected in water contact angles as the permanently damaged silicone samples exhibited super-hydrophilicity and contact angles of below 5° while the activated yet undamaged silicones exhibited contact angles of around 20° [25]. To summarise, surface activation of silicones is possible without permanently altering the mechanical and morphological properties of the silicone but one must very accurately choose the activation parameters and the after-treatment to prevent hydrophobic recovery.

4.2 General Approaches of Silicone Functionalisation with Polysaccharides

In order to reduce CAUTI, antibiotics such as minocyclin, rifampicin, nitrofural, norfloxacin, and gentamicin are commonly used for catheter coatings [5, 26]. Antibiotics significantly reduce microorganism migration along the catheter, which reduces the possibility for the organisms to enter the bladder and cause infections [27, 28]. They are often embedded in a polymer matrix to ensure their slow release. The release duration and therefore the duration of their antimicrobial efficacy can be varied according to their hydrophilic/hydrophobic nature. The hydrophobic ones such as norfloxacin are released slowly and can be used for long-term catheterisation [26]. However, the large drawback of antibiotics' use is the development of bacterial resistance, resulting in inefficiency of the antibiotic, which in some cases can also result in supra-infections. In this case the infected cell gets co-infected with a different strain of the same organism or with another organism, which leads to the resistance of the new organism to antibiotic treatment and has a long-term negative effect on bladder functioning [4, 29]. The microorganism can develop this resistance before the antibiotic can eliminate the bacteria or they evolve to change the antibiotic attack site while others can rapidly pump out the antibiotic [4]. Other conventional antimicrobial substances, such as quaternary ammonium salts, triclosan, chlorohexidine, titanium dioxide and others, have also been tested but have either raised a lot of health issues regarding their possible toxic influence on the human body and the environment or have shown to be inefficient to a wide bacterial spectrum. Therefore, their use is limited to scientific research mostly [30–32]. Alternative surface modification agents for catheters are therefore being developed and used in order to overcome the disadvantages of antibiotics and other conventional chemical agents. Silver coatings are amongst the most promising ones. They can be used alone or in combination with hydrophilic polymers, which can form hydrogels such as polyethylene glycol or polyvinyl alcohol in order to improve the antifouling properties [33–37]. Silver coatings can reduce CAUTI for 20% compared to unmodified silicone catheters. The bacteria develop much less resistance to silver coatings in comparison with antibiotic coatings. The major drawback of silver coatings on the other hand is the possible toxicity of silver to human biological environment [38]. Braydich-Stolle et al. [39] investigated the citotoxicity of silver on mice stem cells and came to the conclusion that silver nanoparticles of concentrations 5 μg/mL $< c < 10$ μg/mL cause necrosis and apoptosis of cells [39]. Yang et al. [40] have reported that the toxicity of silver nanoparticles to *Caenorhabditis elegans* linearly increases with increasing amount of dissolved silver ions, which are directly responsible for the toxicity of the nanoparticles. The nano-particle size on the other hand had no influence on their toxicity [40]. As reported by the US Agency for Toxic Substances and Disease Registry and the European commission for heavy metal waste, the deposition of silver in the environment also contributes to accumulation of heavy ions in nature [41]. Other metals such as gold [36], copper [42], and zinc [43] are also used as coating agents for urethral catheters but find less application then silver. They share their advantages and disadvantages, such as possible toxicity and environmental issues, with silver and therefore the scientific and industrial spheres are searching for new possibilities that will result into improvement of antimicrobial and antifouling properties of silicone catheters. The latest may be achieved with use of natural and biocompatible agents such as polyphenols, polypeptides and mostly polysaccharides, etc. [6, 44, 45]. One of the most promising polysaccharides for medical implant coatings is the second most abundant biopolymer on earth, chitosan (Fig. 4.8). It was mostly studied as an antimicrobial agent on silicone surfaces in combination with other polymers, which exhibit antifouling or lubricant properties, like hyaluronic acid or polyethylene glycol.

In a past study [46], it has been shown that oxygen plasma activation as pretreatment procedure for the application of 1 wt% of chitosan macromolecular solution at pH $= 6$ allows for a better adsorption than untreated silicone. The Fourier-transform infrared (FTIR) spectroscopy results showed that PDMS samples treated with 1% chitosan exhibit more pronounced peaks in the N–H and O–H extensions, as well as in the amide bonding area in comparison to the samples treated by a 0.5% chitosan macromolecular solution. It has been shown that functionalization depends in part on the time of treatment and the plasma treatment site, as well as from the concentration and pH of the chitosan macromolecular solution. In particular, amongst all samples, plasma-activated samples further treated by a 1.0% macromolecular chitosan solution at pH of 6 were outstanding. The contact angles of those functionalized samples were

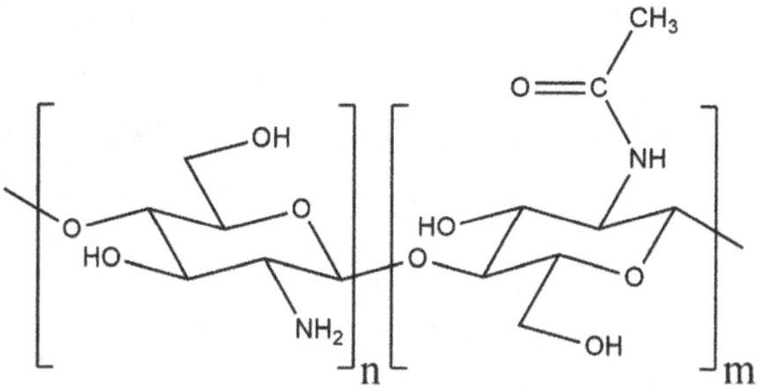

Fig. 4.8 Chemical structure of a chitosan molecule consisting of deacetylated (n) and acetylated units (m)

below 90°. In this way, after the application of chitosan, the hydrophilic character is preserved. Moreover, antimicrobial character, due to the presence of amine groups on the surface of the material was introduced. The inhibition higher than 70% for *Escherichia coli* and *Staphylococcus aureus* was observed.

The obtained hydrophilicity as well as antimicrobial properties are essential for real application of functionalized catheters [46]. In one of the preliminary research studies by the authors of this book, chitosan nanoparticles were prepared by ionic gelation, which is a low-time consuming, simple, and mild reaction without any toxic reagents. The antibiotic amoxicillin was encapsulated into chitosan nanoparticles (Fig. 4.9). Both systems were used as adsorbates for silicone tubes. Functionalized tubes were than analysed regarding elemental composition, morphology and antimicrobial activity. It has been shown that chitosan nanoparticles coating alone or as an antibiotic delivery system, acts as promising functional (hydrophilic and antimicrobial) coating for silicone tubes. The drug was released in a controlled fashion within few hours.

Bongaerts et al. [44] have shown that cross-linked multilayers of hyaluronic acid and chitosan form a lubricating film on silicone, resistant to protein adsorption. A quartz-crystal microbalance (QCM) was used to study the formation of multilayers, whereby the first hyaluronic acid layer was bound to the plasma activated silicone chemically by carbodiimide chemistry and the later chitosan and hyaluronic acid exchanging layers were bound to each other electrostatically (Fig. 4.10).

As can be seen from Fig. 4.10, the frequency (ΔF) of the oscillating crystal gradually decreases with increasing number of bilayers, reaching a final value of $-160\,Hz$ for the 5 bilayer coating. This is an indication that the process of bilayer formation was successful. This is underlined by the dissipation values of the crystal, which are increasing with increasing number of bilayers. Increasing dissipation values reflect an increasing viscoelasticity of the coatings which can later be related to the protein repellent properties [14] of the coated silicone surfaces. In fact, the 5 bilayers of hyaluronic acid and chitosan have proven to reduce human saliva adhesion by a factor 4 when compared to uncoated silicone. The interactions between the saliva and the silicone surfaces were studied by QCM and are shown in Fig. 4.10 (bottom),

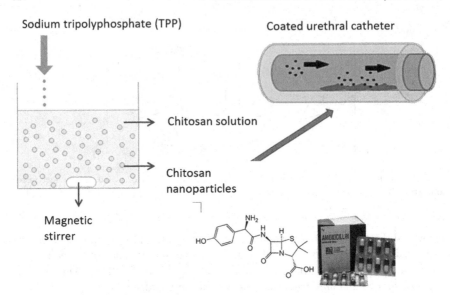

Fig. 4.9 Chitosan nanoparticles as a drug delivery system for silicone tubes

where one can observe that the frequency decreases to around -90 Hz when the saliva deposits on neat silicone films (blank squares and grey circles). In the case of 2 and 5 bilayer coated silicone films the frequency decreases to only -20 Hz, indicating that less saliva is deposited on the surface and that the coatings indeed have a repellent effect [44]. A similar study was successfully conducted previously by Croll et al. [47], where poly(D,L-lactide-co-glycolide) was used as the matrices' substrate instead of silicone. Their concept named "blank slate" describes a surface which is resistant to unspecific protein adsorption but allows to bind bio-functional molecules like extracellular matrix proteins, polysaccharides, or growth factors. This was achieved by a layer-by-layer electrostatic deposition of hyaluronic acid and chitosan which allowed for further covalent binding of collagen IV, which can be used as a template for self-assembley of basement membrane components [47]. Mannan and Pawar [48] have used a polyethylene glycol/polyvinyl alcohol/chitosan slurry to incorporate the aminoglycoside antibiotic gentamicin sulphate and coat it on a silicone material by means of the slurry dipping technique. Briefly, when conducting the slurry dipping technique, one prepares a slurry of the desired coating substance and simply dips the substrate in the slurry for a given amount of time, while the slurry is continuously stirred. The coatings were prepared in the presence and absence of chitosan and the agar diffusion antimicrobial tests have shown that the coatings containing chitosan exhibit a wider inhibition zones which highlighting the importance of chitosan's antimicrobial activity. A coating without gentamicin sulphate was not prepared and submitted to antimicrobial testing and therefore, the antimicrobial activity of chitosan cannot be definitely confirmed. The stability of the coatings was not tested and no information on the interaction of the components in the slurry

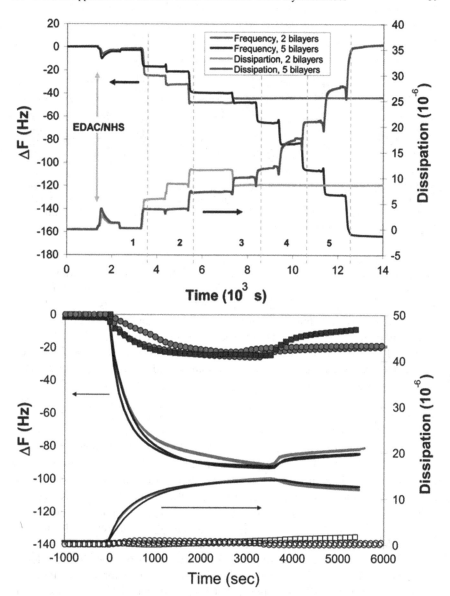

Fig. 4.10 Frequency and dissipation changes of silicone coated gold QCM crystals during the formation of a 2 and 5 bilayer chitosan-hyaluronic acid (top) and during the interactions with saliva (bottom) [44]. This figure was reprinted (adapted) with permission from Bongaerts et al. [44]. Copyright (2009), American Chemical Society

with the silicone elastomer was reported [48]. Kowalczuk et al. [49] have recently used a chitosan hydrogel coating on a commercially available silicon coated latex catheter to further immobilise the antibiotic tosufloxacin on the catheters surface.

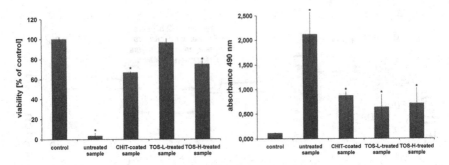

Fig. 4.11 Cytotoxicity evaluation of unmodified and modified silicone catheters by direct-contact methods MTT (3-(4,5-dimethylthiazol-2-yl)-2,5-diphenyltetrazolium bromide assay) (left) and lactate dehydrogenase activity (LDH) (right) [49]. Reprinted from Kowalczuk et al. [49]. Copyright (2015), with permission from Elsevier

The catheters were simply dip coated with the chitosan hydrogel solution, which was later cross-linked with glutaraldehyde. The coated catheter was not submitted to antimicrobial testing or to antifouling testing, only cytotoxicity tests were conducted, which have shown that chitosan improves cell viability and reduces the amount of potentially toxic substances leaching of the catheters base material (Fig. 4.11).

The MTT test results demonstrated that the amount of formazan produced by green monkey kidney (GMK) cells exposed to the unmodified catheters was significantly decreased to 3.3% compared to the control cells cultured with polystyrene samples. Statistical analysis revealed that chitosan coated catheters significantly reduced cell viability to 66.7% and catheter samples treated with a high concentration of tosu-floxacin (TOS-H) reduced cell viability to 74.9% compared to the control. Catheters treated with a low concentration of tosufloxacin (TOS-L) did not induce cytotoxic effect and GMK cell viability was 96.3% compared to the control. LDH test results demonstrated that unmodified silicone latex catheter and all modified samples caused a statistically significant increase in LDH activity released into the medium compared to the control. However, cytotoxic effect of untreated samples was approximately 1.5 times higher than modified catheters [49]. Cell viability evaluation results are presented in Fig. 4.12.

A derivative of chitosan, carboxymethyl chitosan has also gained lots of interest in the last years as an antimicrobial and antifouling substance for silicone implants coating. Wang et al. [50] have shown that a carboxymethyl chitosan coating on silicone can reduce biofilm formation of *Escherichia coli* and *Proteus mirabilis*. Carboxymethyl chitosan has shown to have very good potential to suppress the adhesion of bacteria and it could be expected to have the same effect with proteins. The silicone was first coated with a polydopamine layer to which the carboxymethyl chitosan was later grafted by the Schiff-base reaction. The individual carboxymethyl chitosan layers were later cross-linked with each other by the carbodiimide reaction using 1-ethyl-3-(3-dimethylaminopropyl) and *N*-hydroxisuccinimide (Fig. 4.13).

(a)   (b)

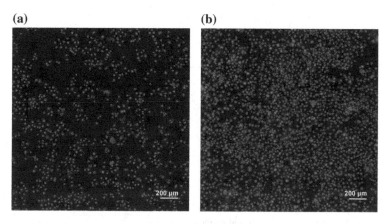

Fig. 4.12   Confocal microscopy images of GMK cells after 24 h of exposure to **a** untreated silicone surface and **b** chitosan coated silicone surface. Blue dots: viable cells, red/violet dots: dead cells [49]. Reprinted from Kowalczuk et al. [49]. Copyright (2015), with permission from Elsevier

The results have shown that fewer bacteria adhere to the carboxymethyl chitosan coated silicone than on the uncoated one (Fig. 4.14). The SEM images show that both *Escherichia coli* and *Proteus mirabilis* cells are mostly distributed singly and separately on the non-cross-linked and cross-linked polydopamine-carboxymethyl chitosan coated silicone (Fig. 4.14c, g, d, h). Contrary, a monolayer of the cells is formed on the pristine (reference) and polydopamine coated silicone (Fig. 4.14a, e, b, f) [50].

The antibacterial test supported these results and has shown that carboxymethyl chitosan coated samples inhibited both *Escherichia coli* as well as *Proteus mirabilis*. In this research, however, protein adsorption was not conducted, as well as no comparison to chitosan coated surfaces was carried out. The excellent antifouling properties of carboxymethyl chitosan were also demonstrated by other authors. Tan et al. [51] have shown that carboxymethylated chitosan inhibits the growth of *Staphylococcus aureus*, *Staphylococcus epidermis*, *Pseudomonas aeruginosa*, and *Escherichia coli* by neutralising the charge of the bacterial membrane, which resulted in flocculation of the bacteria. It inhibits the adhesion of bacteria with an efficiency higher than 90% [51]. In the pursuit to impart protein-repellent properties to silicone surfaces by using a simple green coating process, Bracic et al. [52] have developed a one-step noncovalent coating process using aqueous chitosan and carboxymethyl chitosan particles' dispersions. Firstly, the bio-particles of chitosan and carboxymethyl chitosan were prepared by simple pH adjustment of the solutions. The pH dependent protonation and deprotonation of the amino and carboxyl groups, responsible for the solubility of both polysaccharides, were determined by means of pH-potentiometric titrations. It was shown that chitosan and carboxymethyl chitosan form stable particle dispersions if the pH of the solutions is adjusted close to their points of zero charge. This is at pH 6.5 for chitosan and pH 7 for carboxymethyl chitosan. Interaction studies and the attachment of the bio-particles with thin PDMS films was conducted by means

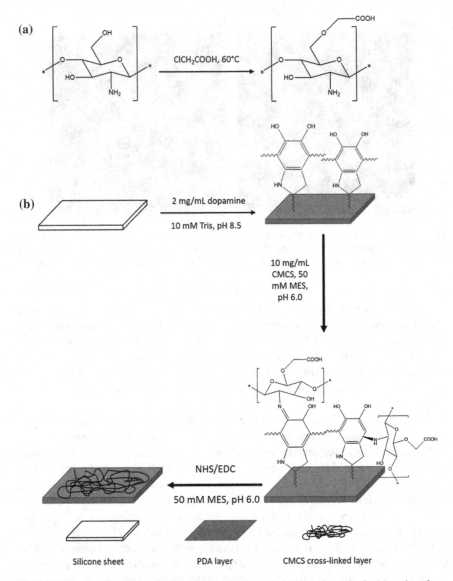

Fig. 4.13 A schematic representation of the carboxymethyl chitosan synthesis (**a**) and surface functionalisation of silicone (**b**) as performed by Wang et al. [50]. Reprinted from Wang et al. [50]. Copyright (2011), with permission from John Wiley and Sons

of QCM-D. Three layers of the bio-particles were deposited on the PDMS films by pumping the chitosan and carboxymethyl chitosan dispersions three times through the QCM-D cell. Intermediate drying of the surfaces was performed in a vacuum

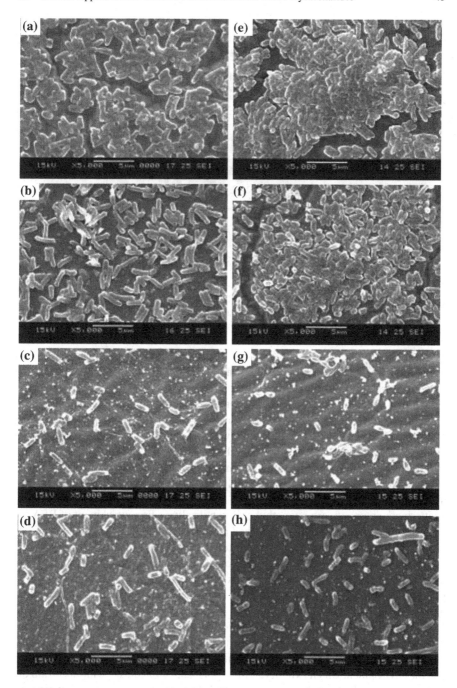

Fig. 4.14 SEM images of *Escherichia coli* (left column) and *Proteus mirabilis* (right column) on pristine silicone (**a**, **e**) and silicone coated with polydopamine (**b**, **f**), carboxymethyl chitosan (**c**, **g**), and crosslinked polydopamine and carboxymethyl chitosan (**d**, **h**) [50]. Reprinted from Wang et al. [50]. Copyright (2011), with permission from John Wiley and Sons

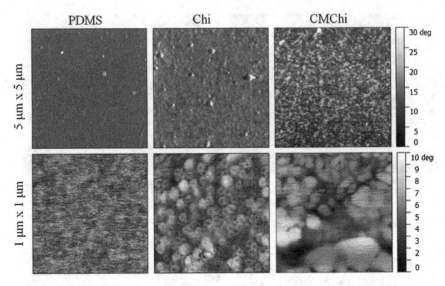

Fig. 4.15 AFM phase images of neat ultra-thin PDMS films and PDMS films coated three times with chitosan and carboxymethyl chitosan bio-particles [52]. Reprinted from Bračič et al. [52]. Copyright (2017), with permission from John Wiley and Sons

oven in order to stabilise each layer. As can be seen in Fig. 4.15, both chitosan (Chi) and carboxymethyl chitosan (CMChi) form a uniform particle coating on the smooth PDMS film when the observed area is 5 μm × 5 μm. Taking a closer look (1 μm × 1 μm), the three chitosan layers (Chi) consist of uniform particle distributed over the whole observed area, while the carboxymethyl chitosan layers (CMChi) consist of larger agglomerates and gaps between the particles can be observed.

The chitosan and carboxymethyl chitosan coated PDMS films were later submitted to interaction studies with bovine serum albumin (BSA) and fibrinogen proteins by means of QCM-D (Fig. 4.16). The protein-repellence test has clearly shown that the zwitterionic nature of carboxymethyl chitosan disallows firm attachment of BSA to its surface as all the protein was detached after rinsing with water. The chitosan on the other hand fails to improve the protein-repellence of the PDMS film as more BSA adsorbs to its surface as does to the PDMS one. The same phenomenon was observed in the case of fibrinogen, where the carboxymethyl chitosan bio-particles improved protein-repellence by more than 3 times, while the chitosan particles improved it only moderately.

The described studies showed that simple and green coating processes using abundant polysaccharides are more than just alternatives to conventional coatings. Non-modified chitosan is an abundant biopolymer, which is harmless for human health and the environment, and it exhibits antimicrobial activity against a wide spectrum of microorganisms. Its main drawback however, is lacking the ability to prevent biofilm formation when coated on an implant's surface. Therefore, it often has to be combined with other hydrophilic or negatively charged polymers to reduce biofilm formation.

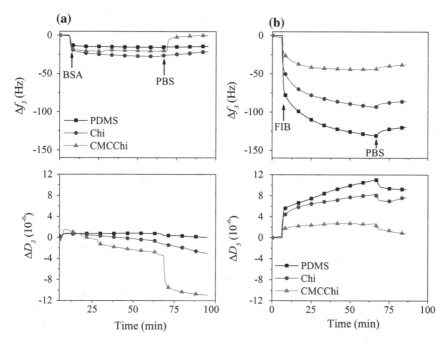

Fig. 4.16 QCM-D frequency and dissipation changes for the adsorption of **a** BSA and **b** fibrinogen, dissolved in phosphate buffer saline (PBS, pH 7.4), onto a native and PDMS surfaces coated with three layers of chitosan and carboxymethyl chitosan [52]. Reprinted from Bračič et al. [52]. Copyright (2017), with permission from John Wiley and Sons

Fig. 4.17 Chemical structure of hyaluronic acid

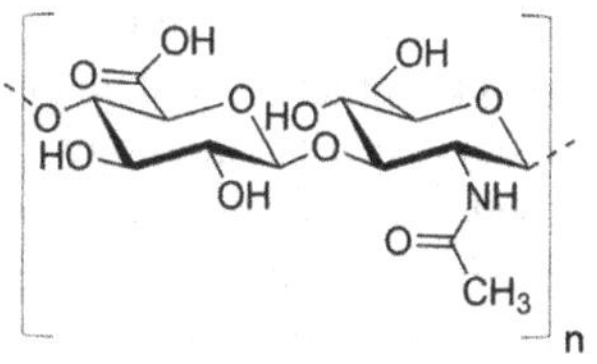

The carboxymethyl chitosan on the other hand provides both antimicrobial activity as well as antifouling properties [52] preventing biofilm formation. All these research results clearly demonstrate very important advantages of carboxymethyl chitosan in comparison to non-modified chitosan. Thus, it is a logical next step in chitosan derivation and underlines the potential of chitosan as a material for bio-functionalisation of medical implants made of silicone.

Besides chitosan and its derivate, hyaluronic acid (Fig. 4.17) has mostly been used in implant coatings for its antifouling and repelling properties. It is naturally found in the human bladder and has been extensively used in studies of antifouling coatings for biomedical applications [15, 44, 53–55].

Patel et al. [55] have investigated coatings of silicone catheters with hyaluronic acid and heparin and compared the growth of epithelial cells of the choroid plexus and

Table 4.1 Time dependent static water contact angles of silicone surfaces coated with N-octadecyltrichlorosilane, heparin, and hyaluronic acid [55]

Time (days)	Static water contact angle (°)		
	Silicone N-octadecyltrichlorosilane	Silicone Heparin	Silicone Hyaluronic acid
0	102.2 ± 1.3	55.3 ± 1.8	55.3 ± 3.9
30	100.8 ± 2.8	55.2 ± 3.2	54.5 ± 2.9

astrocytes with the growth on pristine silicone and hydrophobicaly coated silicone. The silicone was initially activated in a plasma cleaner and subsequently dipped in N-octadecyltrichlorosilane. The hyaluronic acid and heparin were bound to the N-octadecyltrichlorosilane layer by coating them on the silicone from an aqueous solution and exposing them to UV radiation in order to photo-initiate the binding. The stability of the coatings was tested by measuring static water contact angles of the surfaces after they were exposed to simulated physiological conditions for up to 30 days. From Table 4.1 one can clearly observe the contact angle decrease from $102.2 \pm 1.3°$ for the hydrophobic N-octadecyltrichlorosilane coating to $55.3 \pm 1.8°$, and $55.3 \pm 3.9°$ for the hydrophilic coatings heparin and hyaluronic acid, respectively. After 30 days in the saline solution, the contact angles of all coated silicone surfaces remained the same indicating that the polysaccharide coatings were stable. The cell attachment and the cell viability were assessed by trypan blue exclusion and counting with a hemocytometer (Fig. 4.18).

In Fig. 4.18 one can observe the proliferation of astrocyte cells on a polystyrene well plate, pristine silicone, and silicone coated with heparin and hyaluronic acid. The cells attached well and homogeneously to the polystyrene plate. They exhibited the star-like characteristics of glial cells. The pristine silicone surface has shown to provide less fruitful conditions for the astrocytes as only a marginal number of cells were found in this sample ($29.2 \pm 9.6\%$). The hydrophilic heparin and hyaluronic acid coatings increased cell growth on the coated silicone samples. The number of cells for heparin was $70.1 \pm 9.8\%$, and it was $61.6 \pm 7.2\%$ for hyaluronic acid [55].

The interaction between a hyaluronic acid coated silicone catheter and fibrinogen, fibroblast cells as well as corneal epithelial cells was studied by Alauzun et al. [53]. Their experiment showed that a silicone surface modified with hyaluronic acid not only promotes cell proliferation, but also repels proteins at the same time. The protein adhesion was reduced by up to 70% when compared to untreated silicone. The hyaluronic acid was attached to the silicone in a multi-step process (Fig. 4.19). Firstly, a tosylated polethylenglycol layer was grafted onto the silicone surface and amino moieties were introduced to this layer by means of nucleophilic substitution using diethylenetriamine. Finally, the hyaluronic acid was covalently attached to the amino group containing silicone surface using O-(Benzotriazol-1-yl)-N,N,N',N'-tetramethyluronium tetrafluoroborate as the coupling agent [53].

(a)   **(b)**

(c)   **(d)**

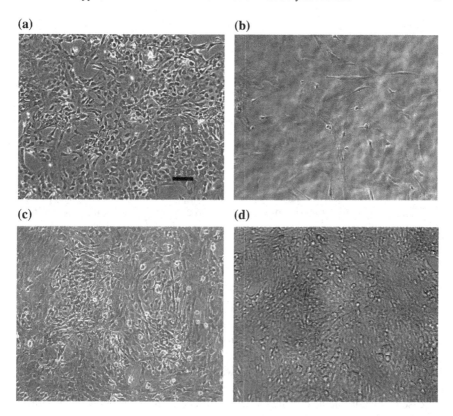

Fig. 4.18 Astrocyte growth on polystyrene well plate (**a**), pristine silicone (**b**), and silicone coated with heparin (**c**), and hyaluronic acid (**d**) [55]. Reprinted from Patel et al. [55]. Copyright (2006), with permission from Elsevier

The static water contact angles measured on the pristine and modified silicone surfaces have revealed that the hydrophilicity of the silicone significantly increases when modified with hyaluronic acid. The contact angles drop from the initial 120° of the pristine silicone to 44° of the hyaluronic acid modified silicone surface, which is in accordance with the results obtained by Patel et al. [55]. A very interesting aspect of this work is the influence of hyaluronic acid to promote cell proliferation and repel protein attachment on the silicone surfaces, which makes it a very promising material for biological applications. The protein interaction studies were performed by labelling fibrinogen with the radioisotope of iodine ^{125}I and immersing the silicone surfaces to a solution of the labelled fibrinogen in human plasma. Later, the radioactivity of the surfaces was determined by a gamma counter and translated to the adsorbed mass of fibrinogen per square centimetre. From the labelling experiment results, it was found out that the mass of adsorbed fibrinogen increased with increasing protein concentration and reached an adsorption plateau at a concentration of around 0.5 mg/mL. 1 μg cm^2 of fibrinogen is adsorbed on pristine silicone

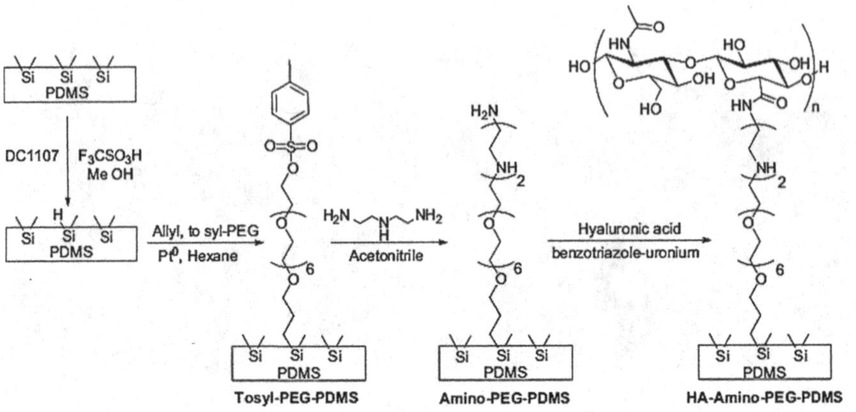

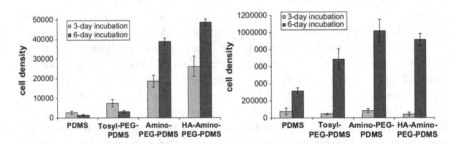

Fig. 4.19 A scheme of the silicone surface funtionalisation path by hyaluronic acid as reported by Alazun et al. [53]. Reprinted from Alazun et al. [53]. Copyright (2010), with permission from Elsevier

at this concentration and only 0.65 μg cm^2 is adsorbed on the silicone surface modified with hyaluronic acid [53]. Optical micrographs of the rat fibroblasts where recorded as well and a general observation from the micrographs is that the area covered with cells on all surfaces increased with increasing number of incubation days. This indicated that the pristine as well as the coated surfaces did not inhibit cell proliferation. Furthermore, the cell-covered areas were significantly bigger in the case of hyaluronic acid coated silicone surfaces. These results were confirmed by cell counts, which are shown in Fig. 4.20.

The antifouling properties of hyaluronic acid on silicone surfaces have also been successfully demonstrated by Wong and Ho [56] as well as by Yue et al. [57], who chemical bonded hyaluronic acid to a surface activated silicone and studied the interaction with albumin and fibrinogen proteins showing a 2.4 and 4.7 fold decrease in adhesion respectively when compared to untreated silicone. Yue et al. [57] applied

Fig. 4.20 Human epithelium cells (left) and fibroblast cells (right) response to pristine (PDMS) and differently modified silicone (Tosyl-PEG-PDMS; Amino-PEG-PDMS; HA-Amino-PEG-PDMS) [53]. Reprinted from Alazun et al. [53]. Copyright (2010), with permission from Elsevier

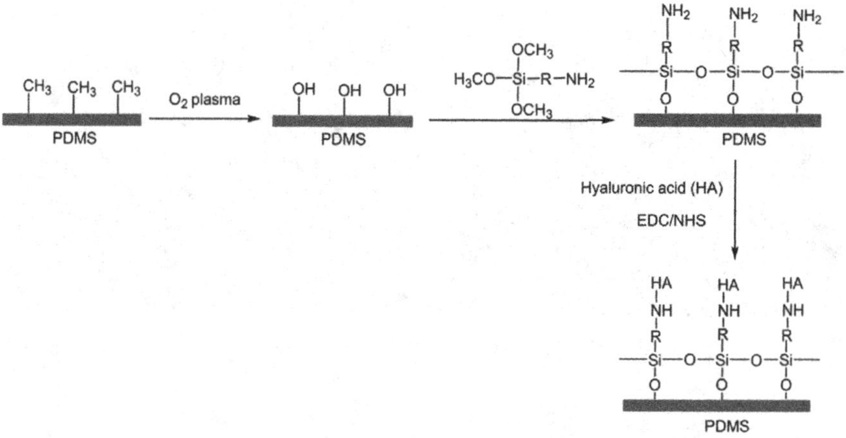

Fig. 4.21 A scheme of the silicone surface modification with hyaluronic acid by carbodiimide crosslinking as reported by Yue et al. [57]. Reprinted from Yue et al. [57]. Copyright (2011), with permission from Elsevier

oxygen plasma treatment of silicone surfaces to introduce hydroxyl moieties on the surface in order to attach an amino group containing silane layer to the silicone which was further used to covalently bind the hyaluronic acid by means of carbodiimide crosslinking (Fig. 4.21). This modification approach is a faster and simpler alternative to that one of Alauzun et al. [53]. Besides the improved protein repellent properties of the hyaluronic acid coating towards albumins and fibrinogen, cell adhesion was studied as well. Neuronal pheochromocytoma cells from rat were used for the cyto-compatibility studies. In Fig. 4.22 one can see that the cell growth on pristine silicone was poorly supported, but the hyaluronic acid-modified one showed moderate cell proliferation at early days in the culture. This indicates to the non-permissive nature of the hyaluronic acid layer to pheochromocytoma cells adhesion. The authors additionally attached collagen to the hyaluronic acid layer by which they even further promoted proliferation of the pheochromocytoma cells [57].

As demonstrated by several authors and described above, one can see that polysaccharides have found many applications in biomedical field, especially as coating agents for silicones. They were used to improve cell viability, repel proteins, and even act antimicrobial. In general, a combination of different polysaccharides or a combination of polysaccharides and antibiotics (in case of antimicrobial properties) is needed to combine several of their intrinsic properties in order to prepare multifunctional coatings. In order to avoid antibiotics' application, mostly because of the overwhelming bacterial resistivity, new approaches were recently developed to replace them with natural antimicrobial substances such as natural, biodegradable surface-active agents (surfactants). Bračič et al. [14] combined these with polysaccharides to develop a novel and single-step coating comprising of biocompatible nanoparticles from a synergistic formulation (HA-MKM) between hyaluronic acid (HA) and a lysine derived natural surfactant (MKM). Considering the structural and

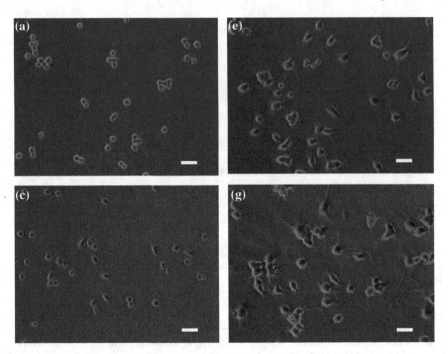

Fig. 4.22 Phase contrast images of the pheochromocytoma cells on hyaluronic acid modified silicone surface at day 1 in growth medium (**a**) and in differentiation medium (**c**), as well as at day 3 (**e**) in differentiation medium and day 6 in differentiation medium (**g**) [57]. Reprinted from Yue et al. [57]. Copyright (2011), with permission from Elsevier

physicochemical diversity of HA and MKM, in addition to the non-destructive surface modification, this study also aimed to understand the interaction of proteins and attachment of microbes at the interfaces of PDMS under dynamic condition, simulating the real conditions to be faced by the catheters, in detail [14]. Francesko et al. [58] used aminocellulose based core-shell nanospheres and hyaluronic acid to form multilayer coatings on silicone catheters able to inhibit bacterial biofilm formation.

In the next chapter, some of the most important findings of these investigations are presented.

4.3 Novel Approaches of Silicone Functionalisation with Polysaccharides

The motivation for the formation of a synergistic formulation between polysaccharides and surfactants by Bracic et al. [14] arose from a previous study by Perez et al. [59], which showed that certain natural surfactants exhibit antimicrobial properties. The MKM surfactant was evaluated for its biological activity. Its haemolytic activity has shown to be lower when compared to conventional cationic surfactants, but its toxicity on the other hand was much lower and its biodegradability higher [59]. Its antimicrobial activity was tested using the minimum inhibitory concentration (MIC) method against a wide variety of microorganisms including *Bacillus cereus* var. mycoides, *Enterococcus hirae, Micrococcus luteus, Staphylococcus aureus, Bacillus subtilis, Staphylococcus epidermis, Mycobacterium phlei, Candida albicans, Klebsiella pneumoniae, Escherichia coli, Salmonella typhimurium, Pseudomonas aeruginosa, Bordetella bronchiseptica, Serratia marcescens*, and *Enterobacter aerogenes.* The results have shown that the MKM exhibits good antimicrobial activity against gram-positive bacteria with MIC values of 64 mg/L. The gram-negative bacteria and the fungi *Candida albicans* have shown to be resistant to the surfactant and their growth could not be inhibited except in the case of *Escherichia coli*, where the MIC value was again 64 mg/L [59].

The preparation of the synergistic formulation between hyaluronic acid and the lysine derived natural surfactant is based on the interactions between ionic surfactants and oppositely charged ionic polymers or as in this case, ionic polysaccharides. These interactions can be described as shown in Fig. 4.23. The diagram shows the change in total surfactant concentration versus the polymer concentration. It is divided in four regions which are all assigned to a specific type of interaction between the polymer and the surfactant. The specific feature of the first region (I) is the very low concentration of the surfactant. It does not induce any interaction with the polymer. Increasing the surfactant concentration leads to the critical association concentration (CAC) which is a surfactant property and defines the concentration at which the surfactant starts binding to the polymer and eventually leads to region II. In this region, the surfactant starts to bind cooperatively to the oppositely charged backbone of the polymer, which induces micellisaton of the surfactant on the polymers backbone as well. Therefore, the CAC commonly lies far below the critical micelle concentration (CMC) of the surfactant [60]. Once all the binding sites of the polymer are occupied by the surfactant, one reaches the second critical association concentration (CAC$_2$), which denotes the end of the binding process. From this point on, adding more surfactant to the system results in an increase of free surfactant molecules in the system which is a specific feature of the third region (III). These free surfactant molecules can act as a stabiliser of the polymer-surfactant complex which can be utilised to form stable colloidal dispersion. The amount of free surfactant molecules will increase until the critical micelle concentration (CMC') is reached. From that point on the fourth region starts (IV). It is important not to confuse the CMC' with the CMC. The CMC is an intrinsic property of the surfactant while the CMC' strongly depends on the type of the polymer the surfactant binds to. It is usually slightly higher than the CMC. The co-existence of a polymer-surfactant complex and free surfactant micelles is a specific feature of region IV [61].

The interactions between the ionic polymer and the oppositely charged surfactant are driven by to major forces. Initially, the surfactant molecules are attracted to the polymer backbone via electrostatic forces. Bringing the surfactant closer to the polymer backbone later allows for the surfactant molecules to bind cooperatively and form micelles on the polymer, thus forming a surfactant-polymer complex

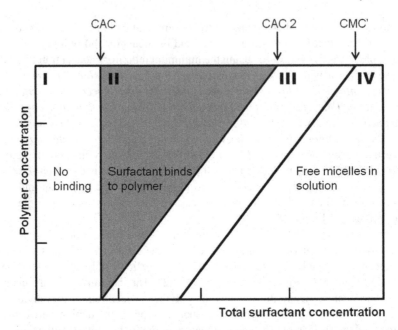

Fig. 4.23 Phase diagram of interactions between oppositely-charged polymer and surfactant [61]. Reprinted from Holmberg et al. [61]. Copyright (2002), with permission from John Wiley and Sons

(Fig. 4.24) [61]. The formation of the complex often results in a precipitate as found by Thalberg and Lindman [62] during their interaction studies between hyaluronic acid and cationic surfactants (alkyltrimethylammonium bromides of various chain lengths). The precipitation starts in region II of the diagram in Fig. 4.23. This is when the electrostatic forces attract the surfactant to the backbone of the polymer and the surfactant micellisation starts as well. The precipitate appears as a pale opalescence dispersion in this region. With the increasing surfactant concentration and the resulting decrease in polymer binding sites, which is specific for region III, the dispersion turns to milky and opaque. The free surfactant molecules stabilise the dispersion in this region until they start forming micelles in region IV. This results in destabilisation of the complex, which sediments and a two-phase system with a bottom white precipitate and a clear dispersion on top can be observed. It was shown that the binding of cationic surfactants to hyaluronic acid is extremely cooperative indicating that binding happens in micelle-like clusters. It was in region III where a stable dispersion of hyaluronic acid and the lysine derived natural surfactant was prepared as well. In this phase, the complex is hydrophobic which enables one to take advantage of the hydrophobic effect to attract the complex with hydrophobic surfaces of medical devices [63].

Bracic et al. characterised the synergistic formulation between HA and the MKM surfactant in detail for its colloidal and physic-chemical properties [63] and deposited it on silicone surfaces [14]. The interaction studies of the synergistic formulation were

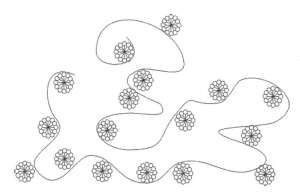

Fig. 4.24 A schematic representation of the surfactant micellisation on the polyion's backbone during the synergistic complex formation [61]. Reprinted from Holmberg et al. [61]. Copyright (2002), with permission from John Wiley and Sons

performed in aqueous media by potentiometric titrations using a surfactant-sensitive electrode and fluorescence microscopy based on pyrene. The CAC of the MKM surfactant in the presence of the sodium salt of HA were determined at different sodium chloride concentrations. During the binding of the surfactant to the backbone of the HA, a reverse binding isotherm was observed which indicated that once the MKM started to bind to HA it induced further binding of free MKM unimers already present in the system, actually reducing the concentration of the free MKM. Actually, the charged and neutral forms of MKM were proposed to explain the anomalous isotherms where even the presence of small amounts of the uncharged MKM strongly influenced the measured electrode response. In such ionic-nonionic surfactant mixtures the amount of non-ionic component cannot be varied independently [63]. The interactions of the HA-MKM formulation with silicone and the protein-repellent properties, as well as antimicrobial activity of such coated silicones were studied by means of QCM-D and a dynamic standard antimicrobial test ASTM E-2149-10 [14].

The synergistic formulation was prepared by mixing aqueous solutions of both the MKM surfactant (positively charged) and the HA (negatively charged) at an exact molar ratio allowing one to prepare stable colloidal dispersions by charge complexation. This yielded bio-particles of plate-like shape and a hydrodynamic diameter of 100–200 nm. The bio-particle dispersions were further used for interactions with silicone thin-films using QCM-D. The bio-particle dispersions were pumped over the silicon thin-films in the closed QCM-D cells using a peristaltic pump. Their interactions with the surface were measured as a function of frequency change of the oscillating QCM-D crystal over a period of time. After the deposition, the surfaces were rinsed with water and dried. The deposition process was repeated three times in order to increase the adsorbed amount of the bio-particles and to create a uniform coating on the silicone [14]. As can be seen from Fig. 4.25, the frequency of the QCM-D crystal dropped from the initial 0 Hz to around −70 Hz indicating that the particles were deposited on the silicone surface. When rinsing with water the

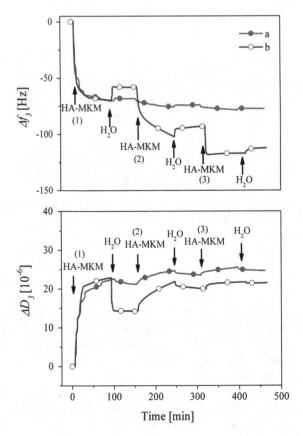

Fig. 4.25 QCM-D frequency (top) and dissipation (bottom) changes for adsorption of HA-MKM bio-particles dispersion on silicone films. No drying (blue) and drying (red) between each adsorption step [14]. Reprinted from Bračič et al. [14]. Published by The Royal Society of Chemistry

frequency increased to around −60 Hz indicating that some of the loosely bound particles were removed. When repeating the adsorption step in wet conditions (Fig. 4.25, blue), one can see that the frequency of the crystal remained almost constant and reached and end value of −75 Hz. The highly hydrophilic particle layer did not allow for further binding as the bonds between the particles in the layer were too loose to be able to bind more particles and allow growth of the coating thickness. This was possible if one stiffened the layer and formed more intermolecular bonds between the polysaccharide molecules. This was achieved by drying of the bio-particles layer after each adsorption step. In this case the next layers could attach to the previous layer without detaching its outermost particles which would lead to desorption. As can be seen from the red curve in Fig. 4.25, the frequency of the crystal gradually decreased with each adsorption step and reached an end value of −110 Hz.

The surface morphology of the bio-particles coated silicone films was characterised by means of atomic force microscopy and confocal laser scanning microscopy as depicted in Fig. 4.26. Randomly distributed bio-particles with voids between them were observed after one layer was adsorbed on the silicone surface. With increasing number of bio-particles layers the fluorescence became more uniform and brighter indicating that more bio-particles are deposited and a more uniform distribution is achieved. This was in good agreement with the QCM-D results. The AFM images revealed an even more interesting deposition profile of the bio-particles on the silicone thin-film. As can be seen in Fig. 4.26. the uncoated PDMS silicone showed a smooth surface with low roughness (Sq: 0.8 nm). After the first adsorption step (a), the inhomogeneous distribution of HA-MKM nanoparticles as well as many large aggregates and a higher surface roughness (Sq: 13.1 nm) were clearly visible. The silicone surface was also not fully covered and larger voids between the bio-particles were observed. After the second and third deposition step, a more organized, densely packed dried bio-particles and a decreased surface roughness (Sq: 10.2–8.1 nm) were noticed [14]. The functionalised silicone thin-films (after three adsorption steps) where then subjected to protein interaction studies by means of QCM-D. This gave an inside on the ability of the coatings to prevent biofilm formation. The functionalised films were mounted in the QCM-D device and aqueous solutions of the proteins bovine serum albumin, fibrinogen, and lysozyme were pumped over the coatings to observe their interactions. As depicted in Fig. 4.27 one can see that the HA-MKM bio-particle coating completely suppressed the attachment of bovine serum albumin as well as lysozyme proteins even after an increase of the protein concentration from 0.1 to 10 mg/mL. This was not the case for the pristine silicone thin-film as the mass of the adsorbed proteins increased with their increasing concentration [14].

As for the fibrinogen, one could observe, some of it attached on the coating, but the adsorbed mass reached values of up to ten times lower than in the case of pristine silicone films. From these results it was concluded that the coating exhibited antifouling properties which arose from the zwitterionic nature of the coating as the HA contains carboxylic and the MKM amino functional groups as well as from the highly hydrophilic hyaluronic acid which has a high water uptake, preventing the protein molecules to deposit on the coated surface [14].

Furthermore, the antimicrobial activity of the silicone, coated with the HA-MKM bio-particles, was tested by means of a dynamic-contact test, which was performed according to the ASTM E-2149-10 standard. The test was performed on urethral catheters made of silicone, which were coated with the bio-particles in accordance with the findings from the interaction studies performed on thin silicone films. The samples were subjected to Gram negative bacteria *Escherichia coli*, *Proteus Mirabilis*, *Pseudomonas aeruginosa*, Gram positive *Staphylococcus aureus*, and the fungi *Candida Albicans*. The number of bacterial colonies was measured before contact with the sample and after it was in contact with the samples for one hour under dynamic conditions. The reduction of bacterial growth was calculated from that colony counts and the results are depicted in Fig. 4.28 [14].

As can be seen from Fig. 4.28, one layer of the particles coating reduced the growth of up to 65% of *the Staphylococcus aureus* colonies. The performance was lowest in the case of *Proteus mirabilis*, where the reduction of growth reached a value of 35%. The antimicrobial activity strongly depended on the mass of the coating. Increasing

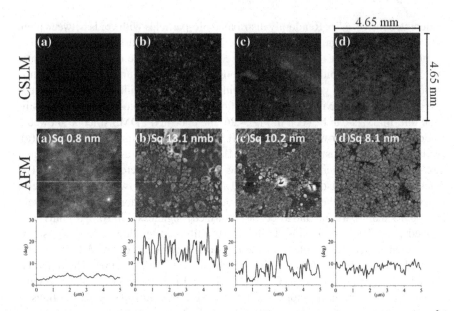

Fig. 4.26 Confocal laser scanning microscopy (CLSM) (top) and AFM phase images ($5 \times 5 \, \mu m^2$, middle) and their respective cross-section profiles (bottom) of uncoated (**a**) and HA-MKM coated PDMS thin-films, **b** first, **c** second, **d** third layer [14]. Reprinted from Bračič et al. [14]. Published by The Royal Society of Chemistry

the number of bio-particles layers showed to increase the reduction of growth as well. The growth of *Staphylococcus aureus*, *Pseudomonas aeruginosa*, and *Escherichia coli* was reduced by over 75% in all cases. The lowest growth reduction of 45% for the three-layer coating was achieved for *Proteus mirabilis*, which showed to be very persistent and growing very fast on the coated surfaces. These results proved that the HA-MKM coating greatly improved the antimicrobial activity of pristine silicone. It showed great potential to be used as a coating for urethral catheters as well as other silicone-based implants, when combining its antimicrobial and antifouling properties [14].

Francesko et al. [58] prepared a coating for silicone urethral catheter based on antimicrobial polycationic nanospheres. The nanospheres were made of aminocellulose and were attached to the silicone surface in a multilayer fashion by exchanging layers of the cationic nanospheres and the anionic HA. The aminocellulose conjugate was firstly sonochemicaly processed in order to form the nanospheres, thus improving its antimicrobial activity compared to the bulk aminocellulose in solution. The multilayer attachment to the silicone was performed by firstly adsorbing a 3-(aminopropyl)triethoxysilane layer on the silicone to allow HA binding. Secondly, a layer of the HA was adsorbed on the surface followed by a layer of the nanospheres. This process was followed on-line by QCM-D and repeated a few times. The bulk nanospheres in solutions were also attached n the same way for comparison purpose. The multilayer adsorption is shown in Fig. 4.29. As can be observed in the QCM-D

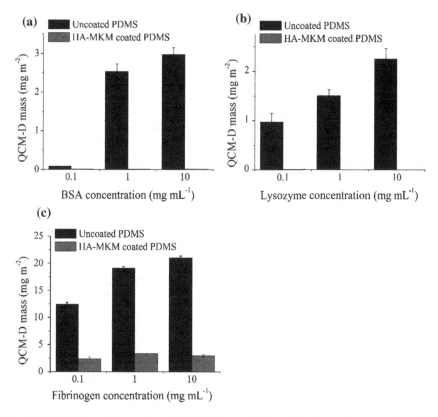

Fig. 4.27   QCM-D end frequencies of HA-MKM coated silicone thin-films after adsorption of the proteins bovine serum albumin, lysozyme, and fibrinogen [14]. Reprinted from Bračič et al. [14]. Published by The Royal Society of Chemistry

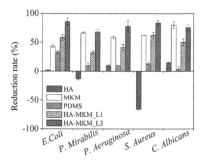

Fig. 4.28   Antimicrobial activity of pristine PDMS silicone and silicone coated with one and three layers of the HA-MKM bio-particles [14]. Reprinted from Bračič et al. [14]. Published by The Royal Society of Chemistry

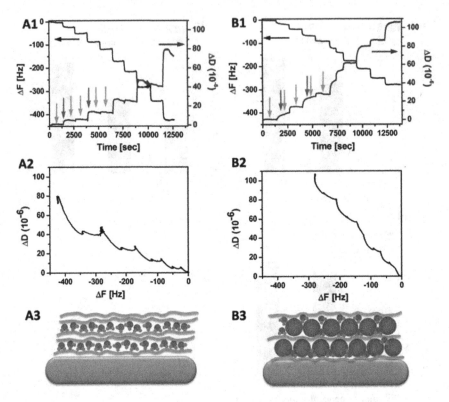

Fig. 4.29 QCM-D frequency and dissipation changes during the build-up of multilayers made of HA and the cationic aminocellulose in solution (left column) and as nanospheres (right column). The coloured arrows represent the introduction of HA (green), water (red), and aminocellulose (blue) in the QCM-D system. A scheme of the multilayer-build up is shown in the bottom row [58]. Reprinted from Francesko et al. [58]. Copyright (2016), with permission from Elsevier

data the frequency of the oscillating crystal decreased exponentially each time one of the components was introduced to the system which indicated that a layer of that compound was successfully adsorbed. One can also observe that the final frequency was higher in the case of the aminocellulose solution (−400 Hz) than in the case of the nanospheres (−300 Hz). The dissipation versus frequency plot also showed that in the case of the nanospheres, less material which is more dissipative is adsorbed in the case of the nanospheres. Thus, a more viscoelastic layer is formed with a higher multilayer thickness. The thickness of the nanospheres multilayer coating was determined to be 141.5 nm, while it was 82.33 nm for the solution coating [58].

The morphology of the coated silicone was observed by means of AFM and SEM (Fig. 4.30). From the AFM images one observed a rough surface in both cases, the multilayer with aminocellulose from solution and in the case of nanospheres. The nanospheres were also distinguishable on the surface of the multilayers as depicted in the SEM images. Nevertheless, the means square roughness of the coating with

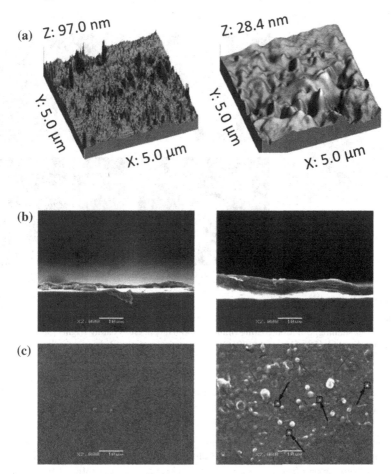

Fig. 4.30 **a** AFM images (5×5 µm), **b** Cross-section, and **c** SEM images of the silicone substrates coated with the HA-aminocellulose multilayers from aminocellulose solution (left) and as aminocellulose nanospheres (right) [58]. Reprinted from Francesko et al. [58]. Copyright (2016), with permission from Elsevier

the nanospheres was smoother (3 nm) in comparison with the coating manufactured from solution (7 nm). The cross-section images also point to the irregularity of the coating thickness. The nanospheres coating was thicker as expected from the QCM-D measurements.

Antimicrobial activity of the multilayer nanosphere coating was determined after attachment to the surface of the silicone substrate. The nanospheres showed a 40% higher antimicrobial effect in comparison to the aminocellulose coating in solution. This also resulted in a higher prevention of *Pseudomonas aeruginosa* biofilm forma-

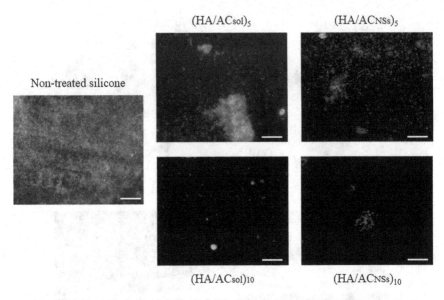

Fig. 4.31 Fluorescence microscopy images (20× magnification) of *Pseudomonas aeruginosa* biofilms on pristine (left) and silicone coated with the HA-aminocellulose multilayers analysed after live/dead staining. Solution coating (middle) and nanospheres coating (right) [58]. Reprinted from Francesko et al. [58]. Copyright (2016), with permission from Elsevier

tion. The stability of the nanosphere coatings was high after exposure to water for 6 and 24 h, but the coatings showed disintegration in the presence of bacteria after 24 h. Nevertheless, the nanosphere coatings successfully inhibited biofilm formation and acted antimicrobial as can be seen in Fig. 4.31.

As can be seen in Fig. 4.31, the authors clearly showed a uniform biofilm formation of *Pseudomonas aeruginosa* pristine silicone while less or no biofilm formation was observed on the coated silicone. Both coatings performed rather equal regarding the biofilm formation while the nanospheres coating clearly acted more antimicrobial as revealed by the red coloured areas in the images, representing dead bacteria on the silicone surface. Moreover, the authors coated silicone Foley catheters with such nanosphere multilayer coatings and showed reduced biofilm formation for up-to seven days, which is considered much longer as a single catheter usage time [58]. In a separated work, the authors attached the aminocellulose nanospheres on silicone surfaces directly, without the HA multilayer formation [64]. In this case the silicone surface was pre-treated with 3-(glycidoxypropyl) trimethoxysilane, which allowed for epoxy/amine curing reaction with the aminocellulose. The shape and size of the attached nanospheres was clearly visualised by means of SEM (Fig. 4.32). The nanospheres were of regular spherical shape and some oval shaped spheres could also be observed. The size of the spheres varied from tens of nm to few μm in diameter [64].

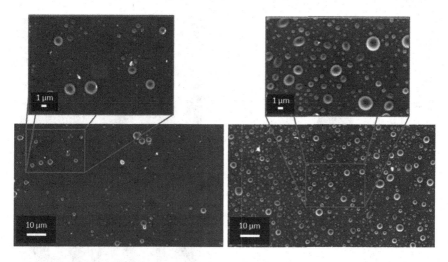

Functionalisation at pH 3   Functionalisation at pH 6

Fig. 4.32 SEM images of the aminocellulose nanospheres attached on silicone [64]. Reprinted from Fernandes et al. [64]. Copyright (2017), with permission from Elsevier

The nanospheres showed high reduction of *Escherichia coli* biofilm formation when coated to silicone surfaces as well as polystyrene surfaces. As can be observed in Fig. 4.33, the nanospheres reduced the biofilm formation of *Escherichia coli* for more than 80% when compared to an untreated epoxy silicone surface.

Coating of silicone catheter was also performed by Bracic et al. [65]. Their studies on chitosan, carboxymethyl chitosan as well as HA-MKM bio-particles activity against protein adhesion and bacteria were extended to silicone catheters as well. In a recently submitted work the authors coated silicone tubes by dip-coating dispersion of chitosan, carboxymethyl chitosan and HA-MKM bio-particles. The process repeated three times in order to reproduce the 3-step adsorption process developed previously on thin silicone films [14, 52]. By doing so the coating was more uniformly distributed on the silicone surface and more mass was attached. The coatings were also tested for their abrasion resistance by using a modified surface abrasion device mimicking the insertion and ejection of a urethral catheter. As can be seen from the confocal laser scanning microscopy images in Fig. 4.34, the silicone tubes are uniformly coated with the chitosan, carboxymethyl chitosan and HA-MKM bio-particles. After one abrasion cycle (forward-insertion, backward-ejection) the coatings are slightly disrupted and removed from the surface, while they are completely removed after a continuous abrasion of ten cycles.

The bio-particle coatings also exhibited slow release when exposed to aqueous environment at pH of 4.5, 7, and 8 for seven days. The released amount of the coatings was determined potentiometricaly and the results showed that at all pH around 50% of the coatings was removed after 10 h. The remaining 50% of the coating remained firmly attached even after seven days of exposure. The pH values on the

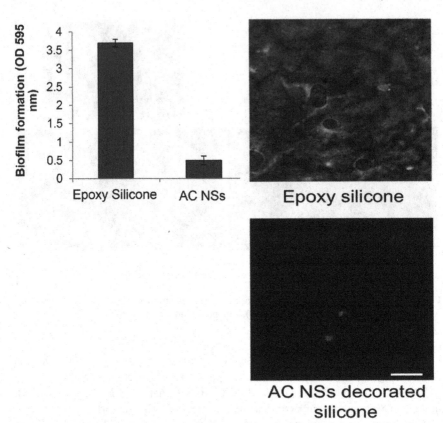

Fig. 4.33 a crystal violet assay of the *Escherichia coli* biofilm formation. **b** Fluorescence microscopy images of the *Escherichia coli* biofilm formation on epoxy silicone and nanosphere coated silicone after live/dead staining [64]. Reprinted from Fernandes et al. [64]. Copyright (2017), with permission from Elsevier

other hand influenced the kinetics of the desorption process. A controlled release could be observed for all bio-particles at pH 7 over the whole period of seven days. This was not the case at pH 4.5 as both types of chitosan began to solubilise and get removed from the surface faster due to protonation of the amino groups. The same was observed in the case of HA-MKM as the MKM amino groups solubilised. At pH 8 the chitosan was barely removed from the surface as the deprotonated amino groups were not soluble, while the carboxylic groups of the carboxymethylated chitosan deprotonated and solubilised the chitosan from the silicone surface. A similar but less exponential desorption effect was observed in the case of HA-MKM. The coated silicone tubes were also tested for their antibacterial properties and their ability to reduce biofilm formation. As can be seen in Fig. 4.35 the coatings significantly improved the reduction of microorganisms when compared to untreated PDMS silicone, reaching reduction rate values of up-to 85% in the case of HA-MKM coating

4.65 mm

7.75 mm

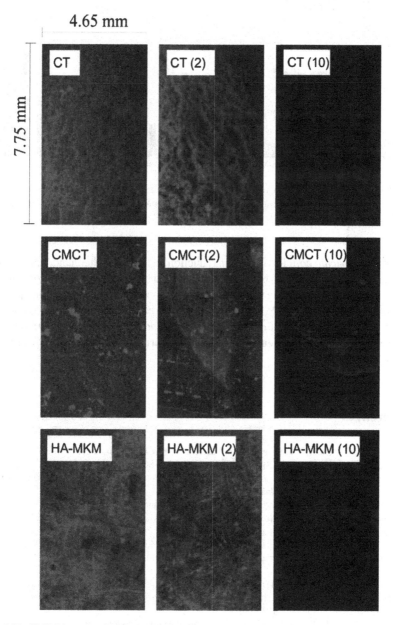

Fig. 4.34 CLSM images of silicone tubes coated with chitosan (CT), carboxymethyl chitosan (CMCT), and HA-MKM bio-particles. Left column: no abrasion, middle column: 1 abrasion cycle, right column: 5 abrasion cycles [65]

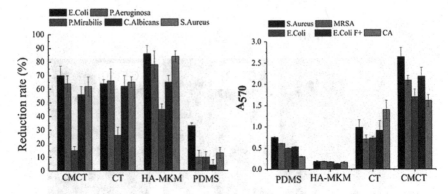

Fig. 4.35 Bioactivity of silicone tubes coated with chitosan (CT), carboxymethyl chitosan (CMCT), and HA-MKM bio-particles. Left: Antimicrobial properties, right: Crystal violet assay of biofilm formation [65]

against *Escherichia coli*. In the biofilm reduction tests on the other hand, the chitosan and carboxymethyl chitosan bio-particles increased biofilm formation in comparison to untreated silicone, making them unsuitable for medical applications. The opposite was observed for the HA-MKM coating, which reduced the biofilm formation of *Staphylococcus aureus*, *Meticilin resistant staphylococcus aureus*, *Escherichia coli*, *and Candida albicans* for more than 50% in comparison to untreated PDMS silicone tubes.

References

1. K. Vasilev, J. Cook, H.J. Griesser, Antibacterial surfaces for biomedical devices. Expert Rev. Med. Devices **6**, 553–567 (2009). https://doi.org/10.1586/erd.09.36
2. M. Bračič, *Surface Modification of Silicone with Polysaccharides for the Development of Antimicrobial Urethral Catheters* (Maribor, 2016)
3. L.E. Nicolle, The chronic indwelling catheter and urinary infection in long-term-care facility residents. Infect. Control Hosp. Epidemiol. **22**, 316–321 (2001). https://doi.org/10.1086/50 1908
4. L. Muzzi-Bjornson, L. Macera, Preventing infection in elders with long-term indwelling urinary catheters. J. Am. Acad. Nurse Pract. **23**, 127–134 (2011). https://doi.org/10.1111/j.1745-759 9.2010.00588.x
5. K. Schumm, T.B.L. Lam, Types of urethral catheters for management of short-term voiding problems in hospitalised adults. Cochrane Database Syst. Rev. 110–121 (2008). https://doi.or g/10.1002/14651858.cd004013.pub3
6. Y. Sh, L. Ys, L. Fh, Y. Jm, C. Ks, Chitosan/poly(vinyl alcohol) blending hydrogel coating improves the surface characteristics of segmented polyurethane urethral catheters. J. Biomed. Mater. Res. B Appl. Biomater. **83**, 340–344 (2007). https://doi.org/10.1002/jbmb
7. E.L. Lawrence, I.G. Turner, Materials for urinary catheters: a review of their history and development in the UK. Med. Eng. Phys. **27**, 443–453 (2005). https://doi.org/10.1016/j.mede ngphy.2004.12.013

8. K. Efimenko, W.E. Wallace, J. Genzer, Surface modification of Sylgard-184 poly(dimethyl siloxane) networks by ultraviolet and ultraviolet/ozone treatment. J. Colloid Interface Sci. **254**, 306–315 (2002). https://doi.org/10.1006/jcis.2002.8594

9. J. Roth, V. Albrecht, M. Nitschke, C. Bellmann, F. Simons, S. Zschoche, S. Michel, C. Luhmann, K. Grundke, B. Voit, Surface functionalization of silicone rubber for permanent adhesion improvement. Langmuir **24**, 12603–12611 (2008). https://doi.org/10.1021/la801970s

10. S. Hemmilä, J.V. Cauich-Rodríguez, J. Kreutzer, P. Kallio, Rapid, simple, and cost-effective treatments to achieve long-term hydrophilic PDMS surfaces. Appl. Surf. Sci. **258**, 9864–9875 (2012). https://doi.org/10.1016/j.apsusc.2012.06.044

11. K. Haji, Y. Zhu, M. Otsubo, C. Honda, Surface modification of silicone rubber after corona exposure. Plasma Process. Polym. **4**, 1075–1080 (2007). https://doi.org/10.1002/ppap.20073 2408

12. A.I. Lopez, A. Kumar, M.R. Planas, Y. Li, T.V. Nguyen, C. Cai, Biofunctionalization of silicone polymers using poly(amidoamine) dendrimers and a mannose derivative for prolonged interference against pathogen colonization. Biomaterials **32**, 4336–4346 (2011). https://doi.or g/10.1016/j.biomaterials.2011.02.056

13. C. Clec'h, C. Schwebel, A. Français, D. Toledano, J.-P. Fosse, M. Garrouste-Orgeas, E. Azoulay, C. Adrie, S. Jamali, A. Descorps-Declere, D. Nakache, J.-F. Timsit, Y. Cohen, Does catheter-associated urinary tract infection increase mortality in critically ill patients? Infect. Control Hosp. Epidemiol. **28**, 1367–1373 (2007). https://doi.org/10.1086/523279

14. M. Bračič, L. Fras-Zemljič, L. Pérez, K. Kogej, K. Stana-Kleinschek, R. Kargl, T. Mohan, Protein-repellent and antimicrobial nanoparticle coatings from hyaluronic acid and a lysine-derived biocompatible surfactant. J. Mater. Chem. B. **5**, 3888–3897 (2017). https://doi.org/10. 1039/C7TB00311K

15. S. Bauer, M.P. Arpa-Sancet, J.A. Finlay, M.E. Callow, J.A. Callow, A. Rosenhahn, Adhesion of marine fouling organisms on hydrophilic and amphiphilic polysaccharides. Langmuir **29**, 4039–4047 (2013). https://doi.org/10.1021/la3038022

16. J. Zhou, J. Yuan, X. Zang, J. Shen, S. Lin, Platelet adhesion and protein adsorption on silicone rubber surface by ozone-induced grafted polymerization with carboxybetaine monomer. Colloids Surf. B Biointerfaces **41**, 55–62 (2005). https://doi.org/10.1016/j.colsurfb.2004.11.006

17. M. Li, K.G. Neoh, L.Q. Xu, R. Wang, E.T. Kang, T. Lau, D.P. Olszyna, E. Chiong, Surface modification of silicone for biomedical applications requiring long-term antibacterial, antifouling, and hemocompatible properties. Langmuir **28**, 16408–16422 (2012). https://doi.org/10.1 021/la303438t

18. A. Oláh, H. Hillborg, G.J. Vancso, Hydrophobic recovery of UV/ozone treated poly(dimethylsiloxane): adhesion studies by contact mechanics and mechanism of surface modification. Appl. Surf. Sci. **239**, 410–423 (2005). https://doi.org/10.1016/j.apsusc.2004.0 6.005

19. E.P.T. De Givenchy, S. Amigoni, C. Martin, G. Andrada, L. Caillier, S. Géribaldi, F. Guittard, Fabrication of superhydrophobic PDMS surfaces by combining acidic treatment and perfluorinated monolayers. Langmuir **25**, 6448–6453 (2009). https://doi.org/10.1021/la900064m

20. D. Maji, S.K. Lahiri, S. Das, Study of hydrophilicity and stability of chemically modified PDMS surface using piranha and KOH solution. Surf. Interface Anal. **44**, 62–69 (2012). https://doi.o rg/10.1002/sia.3770

21. L.F. Zemljič, Z. Peršin, P. Stenius, Improvement of chitosan adsorption onto cellulosic fabrics by plasma treatment. Biomacromolecules **10**, 1181–1187 (2009). https://doi.org/10.1021/bm8 01483s

22. J.L. Fritz, M.J. Owen, Hydrophobic recovery of plasma-treated polydimethylsiloxane. J. Adhes. **54**, 33–45 (1995). https://doi.org/10.1080/00218469508014379

23. M. Bracic, T. Mohan, R. Kargl, T. Griesser, S. Hribernik, S. Kostler, K. Stana-Kleinschek, L. Fras-Zemljic, Preparation of PDMS ultrathin films and patterned surface modification with cellulose. RSC Adv. **4**, 11955–11961 (2014). https://doi.org/10.1039/c3ra47380e

24. D.T. Eddington, J.P. Puccinelli, D.J. Beebe, Thermal aging and reduced hydrophobic recovery of polydimethylsiloxane. Sens. Actuators, B Chem. **114**, 170–172 (2006). https://doi.org/10.1 016/j.snb.2005.04.037

25. S. Béfahy, P. Lipnik, T. Pardoen, C. Nascimento, B. Patris, P. Bertrand, S. Yunus, Thickness and elastic modulus of plasma treated PDMS silica-like surface layer. Langmuir **26**, 3372–3375 (2010). https://doi.org/10.1021/la903154y

26. U.-S. Ha, Y.-H. Cho, Catheter-associated urinary tract infections: new aspects of novel urinary catheters. Int. J. Antimicrob. Agents **28**, 485–490 (2006). https://doi.org/10.1016/j.ijantimica g.2006.08.020

27. R.O. Darouiche, H. Safar, I.I. Raad, In vitro efficacy of antimicrobial-coated bladder catheters in inhibiting bacterial migration along catheter surface. J. Infect. Dis. **176**, 1109–1112 (1997)

28. D. Kowalczuk, G. Ginalska, A. Przekora, The cytotoxicity assessment of the novel latex urinary catheter with prolonged antimicrobial activity. J. Biomed. Mater. Res., Part A **98 A**, 222–228 (2011). https://doi.org/10.1002/jbm.a.33110

29. R. Platt, B.F. Polk, B. Murdock, B. Rosner, Prevention of catheter-associated urinary tract infection: a cost-benefit analysis. Infect. Control Hosp. Epidemiol. **10**, 60–64 (2011)

30. T.A. Gaonkar, L. Caraos, S. Modak, Efficacy of a silicone urinary catheter impregnated with chlorhexidine and triclosan against colonization with *Proteus mirabilis* and other uropathogens. Infect. Control Hosp. Epidemiol. **28**, 596–598 (2007). https://doi.org/10.1086/513449

31. O. Girshevitz, Y. Nitzan, C.N. Sukenik, Solution-deposited amorphous titanium dioxide on silicone rubber: a conformal, crack-free antibacterial coating. Chem. Mater. **20**, 1390–1396 (2008). https://doi.org/10.1021/cm702209r

32. Y. Liu, C. Leng, B. Chisholm, S. Stafslien, P. Majumdar, Z. Chen, Surface structures of PDMS incorporated with quaternary ammonium salts designed for antibiofouling and fouling release applications. Langmuir **29**, 2897–2905 (2013). https://doi.org/10.1021/la304571u

33. M.M. Gabriel, M.S. Mayo, L.L. May, R.B. Simmons, D.G. Ahearn, In vitro evaluation of the efficacy of a silver-coated catheter. Curr. Microbiol. **33**, 1–5 (1996). https://doi.org/10.1007/s 002849900064

34. J. Johnson, P. Roberts, R. Olsen, K. Moyer, W. Stamm, Prevention of catheter associated urinary tract infections with a silver oxide coated urinary catheter: clinical and microbiologic correlates. J. Infect. Dis. **162**, 1145–1150 (1990)

35. H. Kumon, H. Hashimoto, M. Nishimura, K. Monden, N. Ono, Catheter-associated urinary tract infections: impact of catheter materials on their management. Int. J. Antimicrob. Agents **17**, 311–316 (2001). https://doi.org/10.1016/S0924-8579(00)00360-5

36. M. Chung, C. Chin-Chen, Catheter inner surface metal coating by sputtering with microplasma, in *IEEE 35th International Conference on Plasma Science 2008. ICOPS 2008* (2008), p. 1

37. C.Y. Tang, D. zhu Chen, K.Y.Y. Chan, K.M. Chu, P.C. Ng, T.M. Yue, Fabrication of antibacte-rial silicone composite by an antibacterial agent deposition, solution casting and crosslinking technique. Polym. Int. **60**, 1461–1466 (2011). https://doi.org/10.1002/pi.3102

38. P. AshaRani, M.P. Hande, S. Valiyaveettil, Anti-proliferative activity of silver nanoparticles. BMC Cell Biol. **10**, 65 (2009). https://doi.org/10.1186/1471-2121-10-65

39. L. Braydich-Stolle, S. Hussain, J.J. Schlager, M.C. Hofmann, In vitro cytotoxicity of nanopar-ticles in mammalian germline stem cells. Toxicol. Sci. **88**, 412–419 (2005). https://doi.org/10. 1093/toxsci/kfi256

40. X. Yang, A.P. Gondikas, S.M. Marinakos, M. Auffan, J. Liu, H. Hsu-Kim, J.N. Meyer, Mech-anism of silver nanoparticle toxicity is dependent on dissolved silver and surface coating in caenorhabditis elegans. Environ. Sci. Technol. **46**, 1119–1127 (2012). https://doi.org/10.1021/ es202417t

41. D.R. Monteiro, L.F. Gorup, A.S. Takamiya, A.C. Ruvollo-Filho, E.R. de Camargo, D.B. Bar-bosa, The growing importance of materials that prevent microbial adhesion: antimicrobial effect of medical devices containing silver. Int. J. Antimicrob. Agents **34**, 103–110 (2009). https://d oi.org/10.1016/j.ijantimicag.2009.01.017

42. W. Zhang, J. Ji, Y. Zhang, Q. Yan, E.Z. Kurmaev, A. Moewes, J. Zhao, P.K. Chu, Effects of NH_3, O_2, and N_2 co-implantation on Cu out-diffusion and antimicrobial properties of copper plasma-implanted polyethylene. Appl. Surf. Sci. **253**, 8981–8985 (2007). https://doi.org/10.1 016/j.apsusc.2007.05.019

43. D.E. Heskett, *Antimicrobial Urinary Catheter* (2000)

44. J.H.H. Bongaerts, J.J. Cooper-White, J.R. Stokes, Low biofouling chitosan-hyaluronic acid multilayers with ultra-low friction coefficients. Biomacromolecules **10**, 1287–1294 (2009). https://doi.org/10.1021/bm801079a

45. L. Medda, M.F. Casula, M. Monduzzi, A. Salis, Adsorption of lysozyme on hyaluronic acid functionalized SBA-15 mesoporous silica: a possible bioadhesive depot system. Langmuir **30**, 12996–13004 (2014). https://doi.org/10.1021/la503224n

46. B. Polanič, *Površinska obdelava silikonskega materiala* (Maribor, 2016)

47. T.I. Croll, A.J. O'Connor, G.W. Stevens, J.J. Cooper-White, A blank slate? Layer-by-layer deposition of hyaluronic acid and chitosan onto various surfaces. Biomacromolecules **7**, 1610–1622 (2006). https://doi.org/10.1021/bm060044l

48. A. Mannan, S.J. Pawar, Anti-infective coating of gentamicin sulphate encapsulated PEG/PVA/chitosan for prevention of biofilm formation. Int. J. Pharm. Pharm. Sci. **6**, 571–576 (2014)

49. D. Kowalczuk, A. Przekora, G. Ginalska, Biological safety evaluation of the modified urinary catheter. Mater. Sci. Eng., C **49**, 274–280 (2015). https://doi.org/10.1016/j.msec.2015.01.001

50. R. Wang, K.G. Neoh, Z. Shi, E.T. Kang, P.A. Tambyah, E. Chiong, Inhibition of *Escherichia coli* and *Proteus mirabilis* adhesion and biofilm formation on medical grade silicone surface. Biotechnol. Bioeng. **109**, 336–345 (2012). https://doi.org/10.1002/bit.23342

51. Y. Tan, F. Han, S. Ma, W. Yu, Carboxymethyl chitosan prevents formation of broad-spectrum biofilm. Carbohydr. Polym. **84**, 1365–1370 (2011). https://doi.org/10.1016/j.carbpol.2011.0 1.036

52. M. Bračič, T. Mohan, T. Griesser, K. Stana-Kleinschek, S. Strnad, L. Fras-Zemljič, One-step noncovalent surface functionalization of PDMS with chitosan-based bioparticles and their protein-repellent properties. Adv. Mater. Interfaces. **4**, 1–11 (2017). https://doi.org/10.1002/a dmi.201700416

53. J.G. Alauzun, S. Young, R. D'Souza, L. Liu, M.A. Brook, H.D. Sheardown, Biocompatible, hyaluronic acid modified silicone elastomers. Biomaterials **31**, 3471–3478 (2010). https://doi. org/10.1016/j.biomaterials.2010.01.069

54. X. Cao, M.E. Pettit, S.L. Conlan, W. Wagner, A.D. Ho, A.S. Clare, J.A. Callow, M.E. Callow, M. Grunze, A. Rosenhahn, Resistance of polysaccharide coatings to proteins, hematopoietic cells, and marine organisms. Biomacromol **10**, 907–915 (2009). https://doi.org/10.1021/bm8 014208

55. K.R. Patel, H. Tang, W.E. Grever, K.Y. Simon Ng, J. Xiang, R.F. Keep, T. Cao, J.P. McAllister, Evaluation of polymer and self-assembled monolayer-coated silicone surfaces to reduce neural cell growth. Biomaterials **27**, 1519–1526 (2006). https://doi.org/10.1016/j.biomaterials.2005. 08.009

56. I. Wong, C.M. Ho, Surface molecular property modifications for poly (dimethylsiloxane) (PDMS) based microfluidic devices. Microfluid. Nanofluid. **7**, 291–306 (2009). https://doi. org/10.1007/s10404-009-0443-4.Surface

57. Z. Yue, X. Liu, P.J. Molino, G.G. Wallace, Bio-functionalisation of polydimethylsiloxane with hyaluronic acid and hyaluronic acid—collagen conjugate for neural interfacing. Biomaterials **32**, 4714–4724 (2011). https://doi.org/10.1016/j.biomaterials.2011.03.032

58. A. Francesko, M.M. Fernandes, K. Ivanova, S. Amorim, R.L. Reis, I. Pashkuleva, E. Mendoza, A. Pfeifer, T. Heinze, T. Tzanov, Bacteria-responsive multilayer coatings comprising polycationic nanospheres for bacteria biofilm prevention on urinary catheters. Acta Biomater. **33**, 203–212 (2016). https://doi.org/10.1016/j.actbio.2016.01.020

59. A. Colomer, A. Pinazo, M.A. Manresa, M.P. Vinardell, M. Mitjans, M.R. Infante, L. Pérez, Cationic surfactants derived from lysine: effects of their structure and charge type on antimicrobial and hemolytic activities. J. Med. Chem. **54**, 989–1002 (2011). https://doi.org/10.1021/ jm101315k

60. J. Merta, P. Stenius, Interactions between cationic starch and anionic surfactants. Colloid Polym. Sci. **273**, 974–983 (1995). https://doi.org/10.1007/BF00660376
61. K. Holmberg, B. Jönsson, B. Kronberg, B. Lindman, *Surfactants and Polymers in Aqueous Solution*, 2nd edn. (Wiley, West Sussex, 2003). https://doi.org/10.1002/0470856424
62. K. Thalberg, B. Lindman, Interaction between hyaluronan and cationic surfactants. J. Phys. Chem. **93**, 1478–1483 (1989). https://doi.org/10.1021/j100341a058
63. M. Bračič, P. Hansson, L. Pérez, L.F. Zemljič, K. Kogej, Interaction of sodium hyaluronate with a biocompatible cationic surfactant from lysine: a binding study. Langmuir **31**, 12043–12053 (2015). https://doi.org/10.1021/acs.langmuir.5b03548
64. M.M. Fernandes, K. Ivanova, A. Francesko, E. Mendoza, T. Tzanov, Immobilization of antimicrobial core-shell nanospheres onto silicone for prevention of *Escherichia coli* biofilm formation. Process Biochem. **59**, 116–122 (2017). https://doi.org/10.1016/j.procbio.2016.09.011
65. M. Bračič, O. Šauperl, S. Strnad, I. Kosalec, L. Fras Zemljič, Surface modification of silicone with colloidal polysaccharides formulations for the development of antimicrobial urethral catheters. Appl. Surf. Sci. Submitted (2018)

Chapter 5
Conclusions

The microbial biofilm-associated infections caused by medical devices' implantation are still a major healthcare related complication. Amongst them, UTI with a 23% share are the most frequent. Amongst all UTI infections, 80% are CAUTI, as catheters are being used by 10–25% of patients in long-term hospital care with an overall cost of around 395 million EUR/year and 2.550 EUR/treatment.

Once a biofilm has developed on the inside or outside surface of a urinary catheter, the most useful way to eliminate the risk of CAUTI is to give antibiotics with removal of the catheter. There are several problems related to this strategy. First, the pressing clinical question is the duration of antibiotics necessary to treat CAUTI. Second, due to the antibiotic use bacterial resistance was developed and thus healing with antibiotics is limited. Various other prevention strategies generally fall under the headings of different types of catheters, different catheter materials, or alternatives to indwelling urinary catheters. One of the most attractive alternatives to indwelling urinary catheters is to providing antimicrobial, hydrophilic and biofilm inhibiting surface properties those materials. To date, various antimicrobial surface coatings with incorporated conventional antibiotics, triclosan and AgO, have been designed and tested, including the use of biodegradable compounds or hydrogels from PEG or polyvinyl alcohol, polyvinylpyrrolidone, heparin, HA, gendine, and chitosan polymers, to ensure their slow release. Among literature review, many of them describing the use of antibiotics as coatings. Gentamicin, minocycline, rifampicin, etc. were shown to significantly reduce the rate of gram-negative and gram-positive bacteria. However, due to the antibiotic resistance, that is one of the biggest threats to global health, food security, and development today, the use of antibiotics should be reduced. Even more, despite certain promising results, the success rates of the most coatings are still highly variable; some of them may be toxic for the patient and environmentally problematic. Therefore, the scientific and industrial spheres are exploring new possibilities for the improvement of antimicrobial and anti-biofilm-forming properties of catheters surface by using natural and biocompatible agents such as polysaccharides.

M. Bračič et al., *Bioactive Functionalisation of Silicones with Polysaccharides*,
Biobased Polymers, https://doi.org/10.1007/978-3-030-02275-4_5

Unique properties of specific polysaccharides like their interactions with microorganisms, proteins and other biomolecules have opened a door of infinite possibilities to create polysaccharide-based implant coatings. Antimicrobial and antifouling properties are most valued and vital for an implant to provide, when looking from a perspective of implants used in the urethral tract (urethral catheters). Interactions between polysaccharides and materials used for catheter manufacturing, e.g. silicone, are therefore studied in detail in order to create new knowledge, which can be used to prepare durable and effective coatings. New polysaccharides providing these properties have become a main subject of investigation. As demonstrated by several authors in peer reviewed manuscripts, polysaccharides like hyaluronic acid, alginic acid, and some others exhibit a great ability to uptake water. The high content of water plays a pivotal role in preventing protein attachment to silicone catheters coated with such polysaccharides. These polysaccharides mostly contain negatively charged carboxyl groups, which can be combined with positively charged molecules to form zwitterionic complexes that further prevent proteins to attach on the surface of catheters due to electrostatic repulsion forces. The cationic molecule can be either a polysaccharide or of other origin, but cationic polysaccharides are the ones gaining more attention as they often provide antimicrobial activity as well, which is crucial when a catheter is inserted, since bacterial infections cause severe complications during and after the procedure. Very detailed studies were conducted on chitosan as the most known representative of cationic antimicrobial polysaccharides, which have shown that chitosan possesses high antimicrobial activity when used alone or in combination with an anionic polysaccharide as a coating. Besides polysaccharides, other cationic molecules of natural origin are being explored as well. Natural surfactants based on amino acids have proven to have great bioactive properties and some research has already been conducted to prepare synergistic formulations with polysaccharides as well as to test their performance as coatings for silicone catheters. Also, extremely important are the nanotechnologies, thus focus will be given toward the manufacturing of polysaccharides nanoparticle varieties that shall be well established and validated, before scaling up to the surface coating materials where technological sustainability is needed to obtain safety and repeatability of bioactive efficiency.

The success of material functionalization, especially in the case of medical devices, strongly depends on the material's surface properties; i.e. the physics and chemistry of its surface strongly influences its interaction with the chosen environment. The surface of a chemically modified catheter is responsible for the interaction with microorganisms, thus the development and optimization of such surfaces requires a detailed knowledge about the microstructure of its surface. This knowledge is a crucial parameter for enhancement or suppression of functionalized catheter surface efficiency regarding bacteria and fungi inhibition, as well as ability for biofilm performance. Thus, new approaches and models for interaction phenomena understanding are also of a great importance. The future of polysaccharides in medical applications in general is bright. Besides the fields where they have already adapted and are used commercially, new fields are emerging where the potential of polysaccharides has not yet been fully explored. According to a global population distribution analysis by Fisher and Martinez [1], the global population of people aged 65 years and more will be triplicated until 2050. This population represents the majority of patients for catheterisation as well as for other implant materials such as orthopaedic implants surgery.

Importantly, the development of new catheter surfaces will have a positive impact on the aging population, which will reflect in better health as well as a higher quality of life.

Reference

1. A. Fisher, S. Martinez, *Global Population Distribution Analysis* (University of Texas at San Antonio, 2009). http://na.unep.net/globalpop/1-degree/

List of Figure Authorisations

Printed in the United States
By Bookmasters